B·I·O·L·O·G·Y
TEACHER'S
INSTANT
VOCABULARY KIT

MICHAEL L. ROA • DONNELL TINKELENBERG

B·I·O·L·O·G·Y

TEACHER'S

INSTANT

VOCABULARY KIT

With ready-to-use crosswords
& wordsearches for grades 7-12

THE CENTER FOR APPLIED
RESEARCH IN EDUCATION
West Nyack, New York 10995

© 1991 by

THE CENTER FOR APPLIED
RESEARCH IN EDUCATION
West Nyack, New York

10 9 8 7 6 5 4 3 2 1

Library of Congress Cataloging-in-Publication Data

Roa, Michael L., 1946–
    Biology teacher's instant vocabulary kit : with ready-to-use
crosswords & wordsearches for grades 7–12 / Michael L. Roa, Donnell
Tinkelenberg.
        p.   cm.
    ISBN 0-87628-189-7
    1. Biology—Study of teaching (Secondary) 2. Crossword puzzles.
3. Word games. I. Tinkelenberg, Donnell, 1947–   . II. Title.
QH315.R59 1990
574′ .0712—dc20
                                                              90–42597
                                                              CIP

ISBN   0-87628-189-7

**THE CENTER FOR APPLIED
RESEARCH IN EDUCATION**
BUSINESS & PROFESSIONAL DIVISION
A division of Simon & Schuster
West Nyack, New York 10995

Printed in the United States of America

About the Authors

MICHAEL L. ROA (B.A., M.A., San Jose State University) has taught a wide variety of science courses, including biology, physical science, oceanography, environmental science, and general science, to students ranging from fourth grade through college.

Mr. Roa has written articles for *The American Biology Teacher* (journal of the National Association of Biology Teachers), *The Science Teacher* (journal of the American Science Teachers' Association), and *Crucible* (journal of the Science Teachers' Association of Ontario). He has also conducted workshops at local, state, and national teachers' conventions.

Mr. Roa and his students have earned numerous awards in the field of environmental education, including the President's Environmental Youth Award (from the Environmental Protection Agency), and first-place awards for National Energy Education Day projects for the state of California.

DONNELL TINKELENBERG (B.A., Sonoma State University) has taught biology, life science, and physical science at the high school level and health at the junior high school level.

Ms. Tinkelenberg also has a credential in graphic arts. In the late 1960s, she worked as a graphic artist for the University of Maryland's Agricultural Extension Service Office of Information and Publication.

About This Book

Today's students have been exposed to many educational television programs, such as *Sesame Street* and *Nova,* and to educational toys, books, and computer games. Making learning fun, entertaining, and enjoyable is a wonderful thing, but the classroom teacher usually doesn't have the resources (time, money, skills) to produce lessons along the lines of *Sesame Street's* excellent programs. Students, though, sometimes expect school to be as entertaining as Kermit and Big Bird.

While many teachers *are* very entertaining in their presentations, most teachers welcome materials that will help them make their lessons both enjoyable and educationally worthwhile. Many teachers are frustrated by the need to be both educator and entertainer. Students demand that schooling be enjoyable, and the teacher who doesn't provide enjoyable lessons is often met with frustrated and bored students. Thus, you are left with three choices:

1. Spend vast amounts of time and energy trying to provide lessons that are enjoyable, perhaps neglecting content in favor of the method of presentation.
2. Ignore the reality of the student's starting point and expectations by adopting an attitude of "This is school—it's not supposed to be fun."
3. Attempt to provide lessons that the students will enjoy while they learn.

The third choice, of course, is the best—both for the student and the teacher. Therefore, *Biology Teacher's Instant Vocabulary Kit* is intended to provide you with puzzles that can add variety to your life science and biology classes in grades 7–12.

These puzzles, which cover 36 life science and biology topics, are meant to supplement other course materials such as lab guides, reference books, and texts.

For each of the 36 topics included in *Biology Teacher's Instant Vocabulary Kit,* you will find:

- Basic vocabulary lists at the life science level
- Advanced vocabulary lists of terms that would be taught in a biology course
- Definitions of all vocabulary terms
- Crossword puzzle using the basic vocabulary terms
- Crossword puzzle using the advanced vocabulary words

- Vocabulary wordsearch using both the basic and advanced vocabulary terms
- Answer keys to all puzzles (in the back of the book)

The 108 puzzles in this book will help you in several ways:

1. The activities require the students to think and to use terms and concepts correctly.
2. If you are absent, the activities can easily be used by a substitute teacher as self-contained lessons.
3. The puzzles can be reproduced as many times as necessary for different classes.
4. The activities provide models from which you can develop your own puzzles.
5. The vocabulary and concepts are those used and taught in most life science and biology courses and texts. You can easily adapt the materials to your own courses and objectives.

In short, *Biology Teacher's Instant Vocabulary Kit* will help you vary your Biology and Life Science lessons, and inject just the right amount of fun into both teaching and learning!

Michael L. Roa
Donnell Tinkelenberg

How to Use and Modify the Puzzles in This Book

You are encouraged to use the activities in this book as you see fit. The guidelines are just that—guidelines. The puzzles can be used in many creative ways and can be modified to suit your individual style and needs. You are also encouraged to use this book and these puzzles as a basis for creating your own puzzles that are correlated to your own texts, vocabulary, and concepts.

HOW TO USE THE VOCABULARY LISTS

Authors of biology textbooks select the terms they think are the most important for the students using the text to learn. Our comparison of nine of the most commonly-used biology and life science textbooks revealed a wide range of vocabulary that was considered important. We have selected vocabulary from the terms most commonly included in those texts.

The *basic* vocabulary lists are the most basic terms—those most likely to be included in a life science or very basic biology course. The *advanced* vocabulary lists supplement the basic vocabulary with terms that would probably be used in an academic biology course but not in a life science course. The basic vocabulary, of course, would also be included in an academic course.

The inclusion of a term in the vocabulary list does not necessarily mean that it is included in the puzzles. Some terms are used in one activity, some in several, and some not at all. The lists are the "word banks" from which we selected terms to use in the puzzles. If you want to add terms to an activity, or to change the terms, you will find instructions later in this section for doing so.

Since each text is different, it is important that you compare the vocabulary presented in this kit to that used in your classroom text. If terms are included in our activities that are not included in the text being used, you should either delete the term from the activity, "give" that answer to the students, or teach the term. (Each chapter contains definitions for all vocabulary words.)

HOW TO USE THE CROSSWORD PUZZLES

The crossword puzzles can be used to review vocabulary. Teachers have the option of allowing students to refer to the vocabulary lists while working on the puzzles.

Two crossword puzzles are provided for each topic: the first one uses the terms from the basic vocabulary list, and the second uses terms from both the basic and advanced vocabulary lists.

Modification of Crossword Puzzles
1. *Deletion of terms:* Black out spaces for any "unwanted" terms. Be careful not to eliminate letters from intersecting "wanted" words. The clue for that term would have to be removed and the spaces renumbered.
2. *Addition of terms:* Create spaces by using self-adhesive labels over blacked-out squares, or by gluing white paper over them. Lines would then be added to separate the spaces for letters, and clues added and new spaces numbered appropriately. It might be simpler and produce a better-looking crossword puzzle, however, if you use the one provided in this book as a model and copy it, adding additional terms, onto a new graph-paper grid.

HOW TO USE THE WORDSEARCHES

A *Vocabulary Wordsearch* is provided for each topic. Students must first use the clues (definitions) to figure out the terms and then find the terms in the grid.

Modification of Wordsearches
1. *Deletion of terms:* Terms or definitions can easily be deleted from the word list or list of definitions. The grid can be left as is. This would leave unused words in the grid, which is not a problem.
2. *Addition of terms:* Adding or changing terms within the grid would require retyping of the entire grid, unless the added word(s) are obvious. Words might be added within the grid if care is taken not to change any of the other words in the grid. Words can also be easily added to the perimeter of the grid. Of course, the definition list would have to be altered accordingly.
3. *Open-ended wordsearch:* One interesting way to use the wordsearch is to not give the students any clues. The students would be instructed to find and write down as many words pertinent to the chapter as possible in a given time period.
4. *Wordsearch plus:* Another variation is to have the students locate the terms as in paragraph 3 and use them in sentences to demonstrate their knowledge of the meanings.

Contents

B·I·O·L·O·G·Y

TEACHER'S

INSTANT

VOCABULARY KIT

-C — C — C — C — C —

VOCABULARY

Basic

amino acid   molecular formula
carbohydrate   organic compound
chemical energy   peptide bond
enzyme   potential energy
fats   protein
glucose   radiant energy
inorganic compound   structural formula
kinetic energy   substrate

Advanced

(all of the basic vocabulary)
activation energy
dehydration synthesis
disaccharide
fatty acid
glycerol
hydrolysis
lipids
monosaccharide
polypeptide
polysaccharide
radical

DEFINITIONS

activation energy: the energy needed to change potential energy into kinetic energy

amino acid: the nitrogenous building blocks of protein molecules

carbohydrate: organic compound made up of carbon, hydrogen, and oxygen, including sugars, starches, and cellulose

chemical energy: one form of potential energy stored in chemical bonds

dehydration synthesis: the formation of complex organic compounds in which water is given off

disaccharide: sugar formed by the chemical combination of two molecules of simple sugar

enzyme: protein that acts as a catalyst in living things; changes the rate of a chemical reaction without being changed itself

fats: molecules containing fatty acids and glycerol

fatty acid: molecule that combines with glycerol to form lipids

glucose: simple sugar that is the main fuel in both plant and animal cells

glycerol: alcohol molecule that combines with fatty acids to form lipids

hydrolysis: the chemical breakdown of a substance by combination with water

inorganic compound: compound not produced by living things and not containing carbon

kinetic energy: action energy that is doing work or causing change

lipids: fatty substances made up of carbon, hydrogen, and oxygen

molecular formula: notation stating what elements are in a compound and how many atoms of each element are present

monosaccharide: a simple sugar

organic compound: compound that contains carbon

peptide bond: a carbon-nitrogen bond that links amino acids to form polypeptides (proteins)

polypeptide: a large molecule made of amino acids linked by peptide bonds

polysaccharide: large molecule formed by joining monosaccharide molecules through dehydration synthesis

potential energy: stored energy or energy of position

protein: complex chain of amino acids essential to cell structure and function

radiant energy: energy that includes electric waves, radio waves, heat, visible light, and X-rays

radical: the only part of an amino acid molecule that varies

structural formula: diagram showing the atoms in a molecule and how they are bonded

substrate: the substances that enzymes cause to react

Name _____ Date _____

Class _____

ORGANIC CHEMISTRY:
BASIC TERMS CROSSWORD PUZZLE

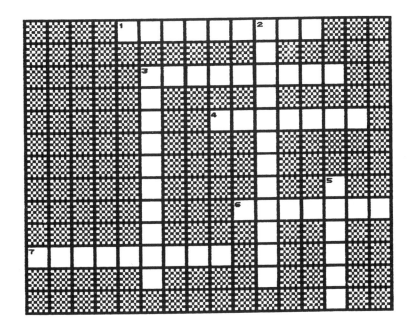

ACROSS

1. One of the nitrogen-containing "building blocks" of proteins (two words)
3. Substance upon which an enzyme acts
4. Chain of amino acids; very important group of compounds
6. Chemical containing carbon; formed by living things
7. Compound not containing carbon

DOWN

2. Organic compound containing carbon, hydrogen, and oxygen; includes sugars and starches
3. A _____ formula is a diagram showing the atoms in a compound and how they are arranged
5. Organic catalyst

Name _____ Date _____

Class _____

ORGANIC CHEMISTRY:
ADVANCED TERMS CROSSWORD PUZZLE

ACROSS

1. Not containing carbon; not formed by living things
3. Polysaccharide made of glucose units organized into branched chains
9. A type of fatty substance
12. A _____ formula is a diagram showing the atoms in a molecule and how they are bonded
13. Formation as of chemical compounds
18. Organic catalyst
19. Organic compound made of carbon, hydrogen, and oxygen; examples are sugars, starch, cellulose
20. _____ acids combine to form fats
21. _____ energy is the energy needed to start a chemical reaction

DOWN

2. Chemical containing carbon; formed by a living thing
4. Something that affects the rate of a chemical reaction without being changed itself
5. Simple sugar
6. A type of carbohydrate; examples are glucose, maltose, lactose
7. One of the simple sugars
8. Very large molecule having a long carbon chain
10. Molecule made up of many sugar molecules combined
11. Made up of many amino acids
14. Nitrogen-containing "building block" of proteins
15. Chemical breakdown of a substance by combination with water
16. Large molecule made up of amino acids subunits linked by peptide bonds
17. _____ synthesis is the formation of complex organic compounds in which water is given off

Name _____ Date _____

Class _____

ORGANIC CHEMISTRY:
VOCABULARY WORDSEARCH

The wordsearch below contains terms related to our study of organic chemistry. The words can be found horizontally in either direction, vertically in either direction, and diagonally in either direction. Clues are given to help you find the words.

```
F L P K L F Y D E S O C U L G X M L G A
J A X W U H O M X A C T I V A T I O N N
E J T U K E D U P E L A C I D A R Z N D
H T G T E D R X N Z Q H Z U B D P H W Y
Q E J Q Y I E J Z A G L Y C E R O L U R
X T P R O R U Y L P D W E J L L L F Y C
H A E W F A D A H N N E C J S I Y Y X A
R R M N Q H W V V I H M R K H A P B T R
K T Y H B C L X I E B N W P S Y E I E B
T S Z Y T C J U Z T V O J Q H I P D D O
S B N O F A W J F O B V Z B V V T I O H
O U E S I S Y L O R D Y H U Q H I C U Y
E S E X V O F M B P M Y N O U Z D A D D
K B N K J N D R O R G A N I C X E O E R
P J R T X O W H C E L L U L O S E N J A
H X V D K M R M V Z S U T V N X H I P T
L W L N A X W L P M V H Z C Q L P M A E
N X W P H T T S Y L A T A C O B U A Z N
```

CLUES

1. Chemistry of carbon or molecules produced by organisms
2. Simple sugar
3. An alcohol that is part of fat molecules
4. Amino acid chain with peptide links
5. Substance on which enzymes act or which enzymes cause to react
6. Nitrogen-containing "building blocks" of proteins (two words)
7. Group of compounds containing carbon, hydrogen, and oxygen
8. Chemical breakdown of a substance by combination with water
9. Chain of amino acids
10. Organic catalyst with protein core
11. _____ energy is the energy needed to change potential energy to kinetic energy, or to get a reaction started
12. _____ acids: along with glycerol, the "building blocks" of fats
13. Large polysaccharide; forms strong fibers in plants
14. Fatty substances
15. Group of atoms that remain together in a chemical change, acting as if they were an atom; for example, OH⁻
16. A chemical that increases the rate at which other chemicals react
17. A simple sugar that is the main "fuel" for cells

5

VOCABULARY

Basic		*Advanced*
cell	multicellular organism	(all of the basic vocabulary)
cell specialization	nuclear membrane	carotene
cell wall	nucleolus	cell fractionization
chlorophyll	nucleus	cell theory
chloroplast	organ	centriole
chromatin	organ system	cytology
chromosome	organelle	grana
cytoplasm	plasma membrane	microfilament
endoplasmic reticulum	plastid	microtubule
eukaryote	prokaryote	middle lamella
Golgi body	ribosome	primary wall
leukoplast	unicellular organism	Matthias Schleiden
lysosome	vacuole	Theodor Schwann
mitochondria		secondary wall
		xanthophyll

© 1991 by Center for Applied Research in Education

DEFINITIONS

carotene: an orange pigment found in chloroplasts

cell: the basic structural and functional unit of all living things

cell fractionization: the process of separating parts of cells with the aid of a centrifuge

cell specialization: the adaptation of a cell for a particular function

cell theory: theory stating that the cell is the unit of structure and function of all living things, and that cells arise only from other cells

cell wall: the outer, rigid layer of plant cells and some protist cells

centriole: organelle in most animal and some plant cells that is involved in the process of cell reproduction.

chlorophyll: the green pigment in plants that traps energy from the sun for use in photosynthesis

chloroplast: plastid containing chlorophyll

chromatin: irregular mass of thin threads of DNA, RNA, and protein that forms the chromosomes

chromosome: contains the genetic code in the form of genes, or units of heredity of DNA

cytology: the study of the structure, organization, and function of cells

cytoplasm: protoplasm outside the nucleus of the cell

endoplasmic reticulum: canal-like membrane system within the cytoplasm

eukaryote: cell with a well-defined nucleus enclosed in a nuclear membrane

Golgi body: organelle made up of a series of closely stacked, flattened sacs that packages substances to be secreted from the cell

grana: membrane in a chloroplast that is made up of proteins, chlorophyll, and lipids

leukoplast: plastid that stores food

lysosome: round organelle containing enzymes, found mainly in the cytoplasm of animal cells

microfilament: protein fibers in a cell's cytoplasm involved in changes in the cell's shape

microtubule: hollow, cylindrical structure in the cytoplasm of animal cells that maintains cell shape and transports substances

middle lamella: outer layer of cell wall that contains a jelly-like substance called pectin

mitochondria: complex oval- and rod-shaped structures in the cytoplasm that function during cellular metabolism

multicellular organism: complete living thing that consists of more than one cell

nuclear membrane: membrane that surrounds the cell nucleus

nucleolus: body within nucleus made up of RNA and protein

nucleus: oval or spherical structure within a cell that contains most of the genetic material necessary for growth and reproduction; the cell's control center

organ: a structure composed of specialized tissues

organ system: a collection of several organs working together to perform a function

organelle: organized structure within a cell that carries out a life function of the cell

plasma membrane: flexible membrane separating the inside of a cell from its surroundings

plastid: organelle that makes or stores food

primary wall: thin layer of cell wall made of cellulose and pectin

prokaryote: cell without a true nucleus; cell in which the nuclear material is not enclosed in a nuclear membrane

ribosome: organelle that makes proteins

Matthias Schleiden: German botanist who proposed that all plants are made of cells

Theodor Schwann: German zoologist who proposed that all animals are made of cells

secondary wall: rigid layer of cell wall that remains after the cell dies

unicellular organism: a complete living thing that is made up of only one cell

vacuole: cavity within the cytoplasm of a cell often filled with food, water, enzymes, and other materials needed by the cell

xanthophyll: a yellow pigment found in plants

Name _____ Date _____

Class _____

CELL STRUCTURE AND FUNCTION: BASIC TERMS CROSSWORD PUZZLE

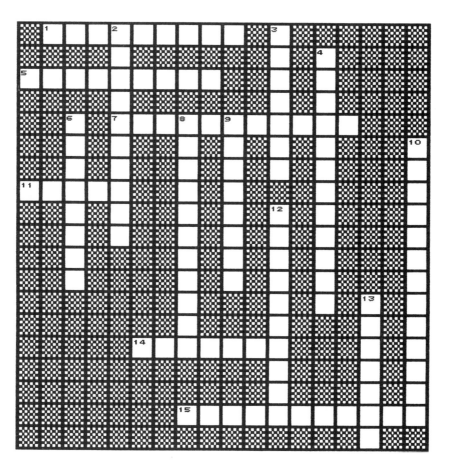

ACROSS

1. Structure within a cell, carries out a life function or process
5. Breaking down of food into small particles that are water soluble
7. Releases energy by the breakdown of glucose molecules
11. The _____ apparatus prepares and "packages" protein secretions in a cell
14. Control center of a cell
15. One-celled

DOWN

2. Process by which the cell takes in digested foods and other chemicals
3. Organelle that stores food or acts as "chemical storehouse"
4. Rod-shaped organelles that are the center of cell respiration
6. Outer, rigid layer of plant cells (two words)
8. Cell that lacks true nucleus and organelles
9. "Protein factory" of cell
10. Made up of more than one cell
12. Rod-shaped structure in the nucleus that actually carries the genetic message
13. Fluid-filled cavity or sac in a cell

Name _____ Date _____

Class _____

CELL STRUCTURE AND FUNCTION: ADVANCED TERMS CROSSWORD PUZZLE

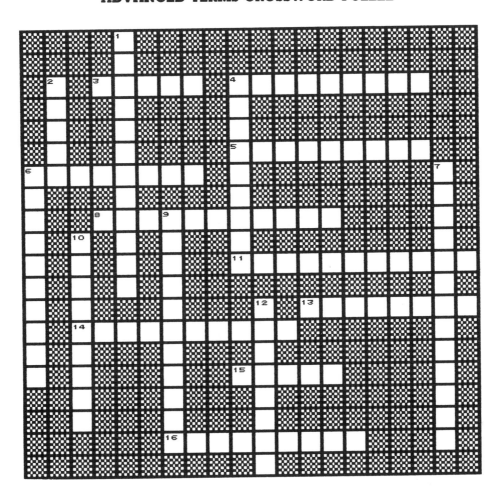

ACROSS

3. Early scientist, used microscope to "discover" and name cells
4. Aids in spindle formation in animal cell division
5. Structure found within a cell that has a specific function
6. Orange pigment found in chloroplasts
8. Process of releasing energy through the breakdown of glucose
11. Long, hollow protein structures that give support to a cell
13. Outer, rigid layer of plant cells (two words)
14. Cell that has nucleus and organelles with membranes
15. _____ apparatus; organelle that prepares and packages protein secretions of the cell
16. RNA, DNA, and proteins in a mass of thin threads in nucleus

DOWN

1. Building of organic molecules by an organism
2. Chlorophyll-containing bodies in the chloroplast
4. Protoplasm or fluid in the cell, but not in the nucleus
6. Rod-shaped structure in nucleus, actually carries the genetic message
7. Outer portion of cell wall, contains pectin
9. Cell without true nucleus or organelles with membranes
10. Small, spherical body within the nucleus that functions in protein synthesis
12. "Protein factory" in cell

CELL STRUCTURE AND FUNCTION:
VOCABULARY WORDSEARCH

The wordsearch below contains terms related to our study of cell structure and function. The words can be found horizontally in either direction, vertically in either direction, and diagonally in either direction. Clues are given to help you find the words.

```
X O N A I R D N O H C O T I M J Q H S J
N L V G R M F B O H N S G Z P O H U D B
I O E X M E Q E W E M O S O M O R H C U
J P X O F I K J J M M Q D A R V B K E G
K P M C M K C O B P R E Y I I W I J G K
L Q I Y L I R R O R N Z B X P E O A J A
J K D T E Q C R O H R O G S E G S A B Z
H L D O U A G R D T S M X L F N Y Q Y O
L A L P K J B C O O U V A G D W N N Y L
E B E L A N Q Y M F F B P L L U T U N Y
U A L A R H O E A F I L U N E E H C O E
A T A S Y F E F B E V L S L T V E L I N
U O M M O A L H T L H X A X E B S E T E
P N E F T B N P O I C U U M R Y I U E T
N R L V I U M A R V Z H N Z E I S S R O
V E L N C Q S N R E L O I R T N E C C R
L L A E T I E H K G J Z X L W R T H X A
Q Q F W T G W Y T E F M K L D Q W T E C
```

© 1991 by Center for Applied Research in Education

CLUES

1. Used a simple microscope to "discover" cells
2. Cell with nucleus and organelles with membranes
3. Fluid within the cell
4. Long, hollow protein structures that give cell support
5. Outermost layer of a plant cell, contains pectin (two words)
6. Building of organic molecules by organisms
7. "Control center" of the cell
8. "Protein factory" of a cell
9. Protein filament in a cell
10. Chlorophyll-containing part of chloroplast
11. Passage of wastes from cell
12. Carries hereditary message; rod-shaped
13. Rod-shaped organelles; centers of cellular respiration
14. Aids in spindle fiber formation
15. Orange pigment

VOCABULARY

Basic		*Advanced*
accessory pigments	glucose	(all of the basic vocabulary)
adenosine	glycolysis	C_4 Pathway
ADP	heterotroph	Calvin (C_3) cycle
aerobic respiration	high-energy bond	carbon dioxide acceptor
alcoholic fermentation	lactic acid fermentation	coenzyme
AMP	light reaction	electron transport chain
anaerobic respiration	low-energy bond	FAD
ATP	oxidation	hydrogen acceptor
autotroph	oxygen debt	NAD
chemosynthesis	phosphate	PGAL
chlorophyll	photosynthesis	pyruvic acid
citric acid cycle	respiration	
dark reaction	visible spectrum	
fermentation	wavelength	

DEFINITIONS

accessory pigments: pigments other than chlorophyll that are found in plants

adenosine: molecule formed from a molecule of adenine and a molecule of ribose

ADP: cellular compound containing one high-energy phosphate bond

aerobic respiration: chemical reaction in which glucose is broken down in the presence of oxygen to produce carbon dioxide, water, and energy

alcoholic fermentation: the anaerobic process that converts glucose to ethyl alcohol

AMP: cellular compound that can form high-energy phosphate bonds to store energy

anaerobic respiration: a process in which a cell converts food into energy in the absence of oxygen

ATP: cellular compound that contains two high-energy phosphate bonds

autotroph: organism that can produce organic molecules from the combination of inorganic molecules

C_4 pathway: stepwise process of carbon fixation that produces glucose faster than the Calvin Cycle

Calvin (C_3) Cycle: pathway that describes the three steps of glucose production during the dark reaction

carbon dioxide acceptor: substance that fixes carbon dioxide to become part of an organic molecule

chemosynthesis: the production of organic compounds through the use of energy produced from chemical reactions

chlorophyll: green plant pigment that traps solar energy for use in photosynthesis

citric acid cycle: a set of steps occurring in the mitochondria during the aerobic stage of cellular respiration

coenzyme: nonprotein molecule that, along with enzymes, acts as a catalyst in chemical reactions

dark reaction: photosynthetic reaction that does not require light

electron transport chain: molecules found in the inner layer of the mitochondria that function as electron carriers

FAD: a coenzyme hydrogen acceptor involved in the citric acid cycle

fermentation: the process by which a cell produces energy from food molecules in the absence of oxygen

glucose: simple sugar produced through the process of photosynthesis

glycolysis: the first stage of respiration that produces pyruvic acid from the breakdown of glucose

heterotroph: organism that is unable to synthesize organic molecules from inorganic molecules

high-energy bond: the breaking of this bond releases energy; joins a phosphate radical to ADP to form ATP

hydrogen acceptor: compound that can bind hydrogen atoms released in chemical reactions and can later release them as needed

lactic acid fermentation: the process that occurs in muscle cells by which glucose is converted to lactic acid

light reaction: reaction in photosynthesis that requires light energy

low-energy bond: bond found in AMP molecule

NAD: a coenzyme that acts as a hydrogen acceptor to produce NADH

oxidation: energy-releasing process that involves the removal of hydrogen from a substance

oxygen debt: state caused by excess lactic acid building up in muscle tissues

PGAL: product of photosynthesis that is the last step before the production of glucose

phosphate: molecule that is bonded to adenosine in ADP, AMP, and ATP

photosynthesis: the process in which carbon dioxide and water are used to form glucose in the presence of sunlight

pyruvic acid: molecule produced by glycolysis

respiration: the process in which glucose is broken down to release energy

visible spectrum: wavelengths that make up white light that can be seen by humans

wavelength: the distance between corresponding points on two successive waves

Name _____ Date _____

Class _____

ENERGY, RESPIRATION, AND PHOTOSYNTHESIS:
BASIC TERMS CROSSWORD PUZZLE

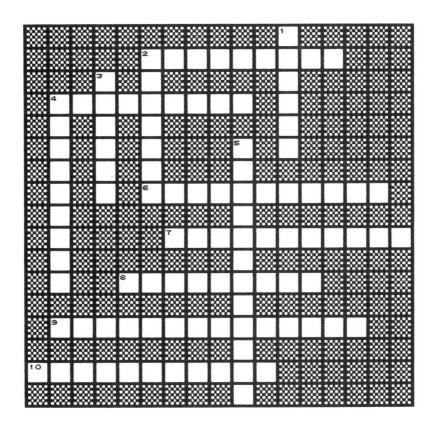

ACROSS

2. Adenine plus ribose
4. Without oxygen
6. Green pigment; photosynthetic
7. Unable to make organic molecules from inorganic substances
8. Removal of hydrogen; energy releasing reaction
9. Glucose formation from water and carbon dioxide
10. Energy released from glucose

DOWN

1. _____ acid cycle occurs in the mitochondria, aerobic stage
2. Requiring oxygen
3. _____ acid fermentation occurs in the muscle cells
4. Organism that can produce organic molecules from inorganic material
5. Releases energy from food without oxygen

Name _____ Date _____

Class _____

ENERGY, RESPIRATION, AND PHOTOSYNTHESIS: ADVANCED TERMS CROSSWORD PUZZLE

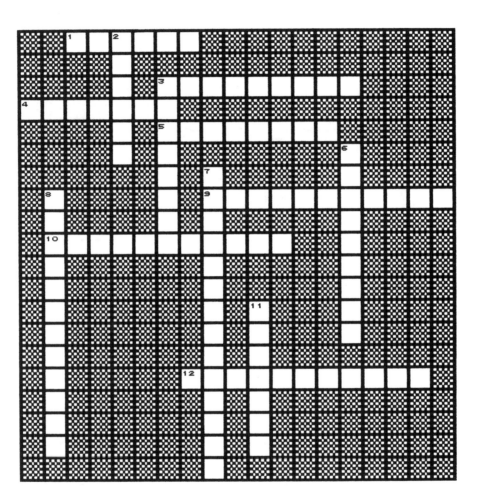

ACROSS

1. _____ acid is formed in the muscle cells from glucose
3. Without oxygen
4. Molecule produced by glycolysis; _____ acid
5. Nonprotein catalyst
9. Unable to produce organic molecules from inorganic material
10. Glucose broken down to release energy
12. Green pigment, photosynthetic

DOWN

2. _____ Cycle; three steps of glucose production in dark reaction
3. Hydrogen _____ can bind to hydrogen atoms and later release them
6. Can produce organic molecules from inorganic material
7. Produces glucose from water and carbon dioxide in sunlight
8. Energy released from food without oxygen
11. Requiring oxygen

ENERGY, RESPIRATION, AND PHOTOSYNTHESIS:
VOCABULARY WORDSEARCH

The wordsearch below contains terms related to our study of energy, respiration, and photosynthesis. The words can be found horizontally in either direction, vertically in either direction, and diagonally in either direction. Clues are given to help you find the words.

```
B K X U H A H Q D N S H A H X H R G
P G K B O S E L G N A F Y O L Y P A
J R U E X L T E B L A G P N Z O Y J
E P U Q N F E R A Q Y R H D A D G V
X Y H G N B R Y N G Q Z T V D K R Y
R R P C P Q O C A E T A H P S O H P
O U N R H W T P E P O E K E I D O S
Z V O T O X R P R K C R Z W P Z A O
Y I I L T L O V O U F O D L T E O F
A C T L O P P I B S I B S U A A D B
D K A Y S I H K I V G I A V V U D P
E N R H Y H I J C H F C D J H Y L N
N Q I P N Z C C I H P O R T O T U A
O R P O T J N O I T A T N E M R E F
S D S R H E V B Y L J O W O A D P R
I Y E O E P H D M F V T Y F X N H S
N X R L S Z N A D A L K O P L U H S
E H W H I Z K U E M Y Z N E O C Y H
A H P C S A Q S I S Y L O C Y L G R
O Z X I W L F I I O H R P V T K T B
```

CLUES

1. Process that releases energy from glucose breakdown
2. Adjective for organism that is unable to synthesize organic molecules from inorganic molecules
3. Adenine plus ribose
4. Coenzyme, acts as hydrogen acceptor
5. Without oxygen
6. Process of releasing energy in absence of oxygen
7. Formation of glucose from carbon dioxide and water
8. Initials for compound with two high energy phosphate bonds
9. Molecule bonded to adenosine in ADP and ATP
10. Produces pyruvic acid from glucose

11. Nonprotein catalyst
12. Initials for last step of photosynthesis before glucose production
13. Adjective for organism that can produce organic molecules from inorganic molecules
14. Adenosine diphosphate
15. Requiring oxygen
16. _____ acid is produced by glycolysis
17. Initials for coenzyme hydrogen acceptor in citric acid cycle
18. Pigment used in photosynthesis

VOCABULARY

adenine
anticodon
base
base pairing
codon
complementary
Francis Crick
cytosine
deoxyribose
DNA
double helix
guanine
m-RNA

nucleic acid
nucleotide
protein synthesis
replication
ribose
ribosomal RNA
RNA
thymine
transcription
triplet
t-RNA
uracil
James D. Watson

(all of the basic vocabulary)
Rosalind Franklin
purines
pyrimidines

DEFINITIONS

adenine: one of the four bases in nucleic acids; it pairs with thymine or uracil

anticodon: exposed base triplet at one end of a t-RNA molecule

base: part of a nucleic acid molecule; can be adenine, cytosine, thymine, guanine, or uracil

base pairing: the process in which complementary base pairs match up during replication, transcription, or translation

codon: group of three nucleotides that codes for a specific amino acid

complementary: word describing a pair of bases that can bond together

Francis Crick: British biophysicist who was one of two men to propose a model for the DNA molecule

cytosine: one of the four bases in nucleic acids; it pairs with guanine

deoxyribose: a sugar that contains one less oxygen atom than ribose

DNA: deoxyribonucleic acid; a molecule consisting of alternate five-carbon sugar units and phosphate molecules connected by nitrogenous bases

double helix: the twisted, ladder-like shape of a DNA molecule

Rosalind Franklin: the British chemist who described DNA as a spiral molecule with bases on the inside of the molecule

guanine: one of the bases in nucleic acids; it pairs with cytosine

m-RNA: messenger RNA; the type of RNA that carries genetic information from the nucleus to the ribosomes

nucleic acid: long chain of many nucleotides chemically bound together

nucleotide: organic compound consisting of a pentose sugar, phosphoric acid, and a nitrogenous base

protein synthesis: the bonding of amino acids to produce protein molecules

purines: class of organic molecules that includes adenine and guanine

pyrimidines: class of organic molecules that includes thymine and cytosine

replication: process in which a DNA molecule builds an exact copy of itself

ribose: sugar in RNA that contains one more oxygen atom than a DNA molecule

ribosomal RNA: type of RNA found in ribosomes

RNA: single-stranded nucleic acid in which the sugar is ribose

thymine: one of four bases in DNA; it pairs with adenine

transcription: the process that transfers information from DNA to RNA

triplet: a name used to describe three bases that code for a specific amino acid

t-RNA: RNA molecule that carries amino acids to the ribosomes for use in protein synthesis

uracil: a base found in RNA that bonds to adenine

James D. Watson: American biologist who was one of two men who proposed a model for the DNA molecule

Name _____ Date _____

Class _____

NUCLEIC ACIDS AND PROTEIN SYNTHESIS:

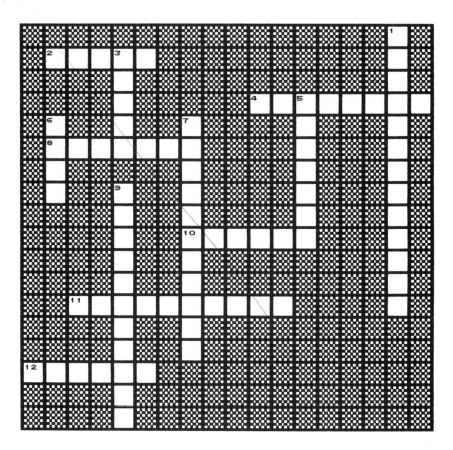

ACROSS

2. One of the scientists who developed the double helix model
4. DNA base, pairs with guanine
8. DNA base, pairs with thymine
10. Sugar in RNA
11. Sugar plus phosphate group plus base
12. Pairs with adenine in RNA

DOWN

1. Process of transferring code from DNA to m-RNA
3. Three bases; represent an amino acid
5. Pairs with adenine in DNA
6. Adenine, guanine, thymine, and cytosine are each an example of a _____ .
7. Sugar in DNA
9. Duplication of a DNA molecule

Name _____ Date _____

Class _____

NUCLEIC ACIDS AND PROTEIN SYNTHESIS:

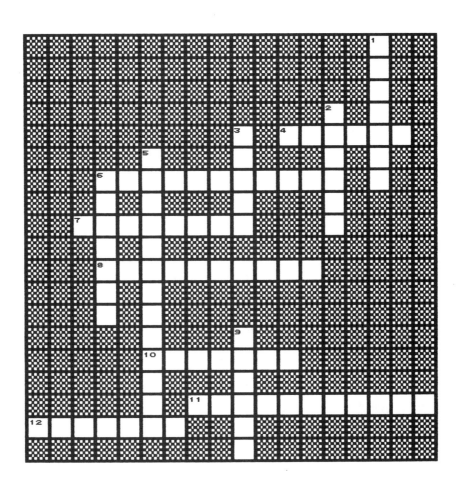

ACROSS

4. Base found in RNA, not in DNA
6. Group of molecules that includes thymine and cytosine
7. One of the scientists who helped develop the double helix model
8. Sugar plus base plus phosphate
10. Combines with adenine in DNA
11. Sugar in DNA
12. Combines with thymine in DNA

DOWN

1. Combines with cytosine in DNA
2. One of the scientists who helped develop the double helix model
3. Three nucleotides that code for a particular amino acid
5. Transferring information from DNA to RNA
6. Group of molecules that includes adenine and guanine
9. Sugar in RNA

NUCLEIC ACIDS AND PROTEIN SYNTHESIS:
VOCABULARY WORDSEARCH

The wordsearch below contains terms related to our study of nucleic acids and protein synthesis. The words can be found horizontally in either direction, vertically in either direction, and diagonally in either direction. Clues are given to help you find the words.

```
Y M E W S Z T X E X I Z M M N B N W
T L L X I V O Y T Z D N R R N A D O
T Q B E Y E Q E D I T O E L C U N C
L U O E G Q C C I E T E N I N E D A
Z X D S F R K O R F H O D D N A V T
F M H O D R R D X B Y V L S C L S X
Q N B B W B P O I D M S B Q D V V C
W Q R I K H K N Q I I B E O J Y M N
T E F R F E B T K P N R U E R N D E
P N C Y C H H O K K E B P N C N L N
I I B X N I B A P P L O E I Y O S I
R N T O T N T H Z E E S E D T I M R
R A U E N J G R H P A S C I O T B U
V U Q D D R N E X B O D G M S A K P
A G C T C U L K G B T Z L I I C N W
V M Y P K I G M I A M U A R N I E Q
S B R R X B T R S D R Z E Y E L S O
X X C N Z T R N A H T M G P F P W Q
E L I C A R U H I I X V G W T E O G
Y N C D W F M D F Q W Y X K B R A M
```

© 1991 by Center for Applied Research in Education

CLUES

1. Deoxyribonucleic acid
2. Sugar in RNA
3. Shape of DNA molecule (two words)
4. Pairs with cytosine
5. Bonds to adenine in RNA
6. Carries genetic information from nucleus to ribosomes
7. Single-stranded nucleic acid, ribose sugar
8. Adenine, cytosine, thymine, guanine, or uracil
9. Adenine or guanine
10. Thymine or cytosine
11. Bonds with guanine
12. Brings amino acids to ribosomes
13. Sugar in DNA
14. Sugar plus phosphate plus base
15. Pairs with thymine or uracil
16. Pairs with adenine
17. Building of copy of DNA molecule
18. Three nucleotides that code for a specific amino acid

VOCABULARY

Basic

anaphase
asexual reproduction
aster
astral rays
binary fission
budding
cell plate
centriole
centromere
centrosome
chromatid
chromatin
chromosome
cleavage furrow
clone
diploid (2N)
egg
fertilization
fission
gamete

gene
haploid (1N)
heredity
homologous chromosomes
interphase
meiosis
metaphase
mitosis
polar bodies
prophase
sexual reproduction
sperm
spindle fibers
spore
telophase
tetrad
trait
vegetative propagation
zygote

Advanced

(all of the basic vocabulary)
synapsis

DEFINITIONS

anaphase: phase in mitosis in which the chromatids separate and move to opposite poles of the spindle

asexual reproduction: reproduction that does not involve the fusion of two haploid nuclei

aster: centriole with astral rays

astral rays: protein fibers that form around the centriole during prophase

binary fission: a form of asexual reproduction in which one cell divides into two cells of equal size

budding: the growth of a small reproductive fragment from a larger parent organism

cell plate: in plant cells, structure that forms between two daughter cells during telophase of mitosis

centriole: in animal cells, organelle that functions in cell division and aids in spindle formation

centromere: place on a chromosome where chromatids are held together before separating

centrosome: in animal cells, small oval area just outside the nuclear membrane that contains one or two centrioles

chromatid: strand of replicated DNA formed during prophase

chromatin: threads of DNA, RNA, and proteins that form the chromosomes

chromosome: rod-shaped structure located in the cell nucleus that carries the hereditary message

cleavage furrow: in animal cells, the pinching inward of the cell membrane to form separate cells during telophase

clone: individual organism produced by asexual reproduction

diploid (2N): having a complete set of chromosome pairs

egg: the female reproductive cell

fertilization: the union of two unlike gametes

fission: form of asexual reproduction in which one organism divides into two or more organisms

gamete: haploid reproductive cell

gene: segment of DNA that controls a hereditary trait

haploid (1N): having only one chromosome of each homologous pair

heredity: the transmission of traits from parent to offspring

homologous chromosomes: chromosomes that have similar and paired genes

interphase: the period of time between cell divisions

meiosis: cell division in which the number of chromosomes in daughter cells is reduced to one-half the number in the parent cell

metaphase: the second phase of mitosis in which chromosomes line up at the cell equator

mitosis: the division of chromosomes into two identical sets

polar bodies: in animals, the three haploid cells that die during the development of the egg

prophase: the phase of mitosis in which the chromosomes contract and the spindle fibers form

sexual reproduction: reproduction that involves the fusion of two haploid cells to form a diploid zygote

sperm: the male reproductive cell

spindle fibers: structure with which the chromosomes are associated during mitosis and meiosis

spore: asexual reproductive cell

synapsis: the pairing of homologous chromosomes during meiosis

telophase: the last phase of mitosis in which the cytoplasm divides and the nuclei reappear

tetrad: a group of four chromatids

trait: characteristic passed from parent to offspring

vegetative propagation: a form of asexual reproduction in plants

zygote: a fertilized egg

Name _____ Date _____

Class _____

CELL REPRODUCTION AND GROWTH:
BASIC TERMS CROSSWORD PUZZLE

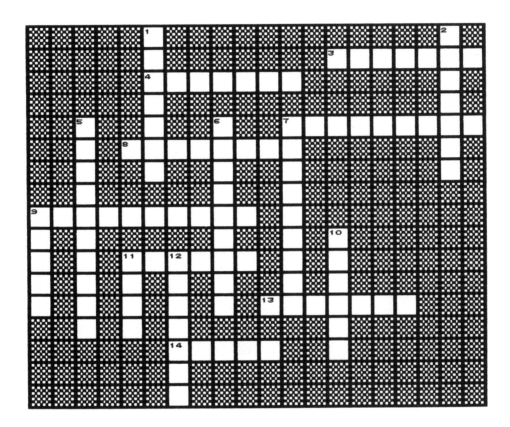

ACROSS

3. Cell with only one of each pair of chromosomes
4. Cell with a complete set of paired chromosomes
7. Strand of replicated DNA, formed during prophase
8. Phase in which chromatids separate and move
9. Carrier of hereditary information
11. Reproductive cell
13. Cell division resulting in haploid cells
14. Male gamete

DOWN

1. A form of asexual reproduction
2. Splitting
5. Period between cell divisions
6. A complex of nucleic acids and proteins found in the nucleus
7. Organelle found in animal cells; aids in spindle formation
9. Exact copy
10. Fertilized egg
11. Segment of DNA that controls a particular trait; unit of heredity
12. Cell division that results in diploid cells

CELL REPRODUCTION AND GROWTH:
ADVANCED TERMS CROSSWORD PUZZLE

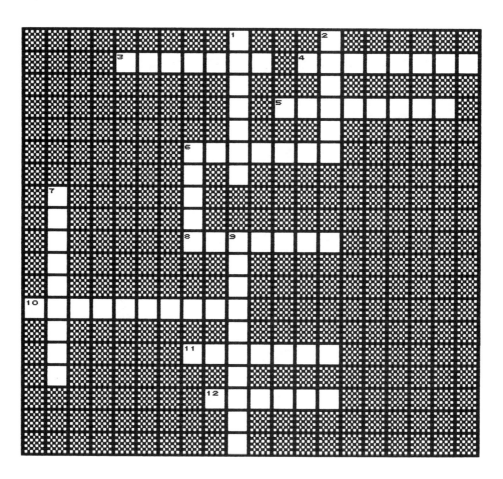

ACROSS

3. 2N
4. Close contact of homologous chromosomes during meiosis
5. Mitosis stage; chromosomes contract and spindles form
6. Fine "threads" between poles of the nucleus in mitosis
8. Results in haploid cells
10. Rod-shaped carrier of hereditary information
11. 1N
12. Sex cell, haploid

DOWN

1. Cell reproduction, results in diploid cells
2. Fertilized egg
6. Male gamete
7. Last phase of mitosis in which the cytoplasm divides
9. Period between cell divisions

Name _____ Date _____

Class _____

CELL REPRODUCTION AND GROWTH: VOCABULARY WORDSEARCH

The wordsearch below contains terms related to our study of cell reproduction and growth. The words can be found horizontally in either direction, vertically in either direction, and diagonally in either direction. Clues are given to help you find the words.

```
F L K G K Q D D O E T R E T S A R F
W K R I W E E O T I Y T I D E R E H
U K K K X L U F U F Z G P V P W A B
N K U Z E D H W K G A Y S R Y V H Y
X D T N S N H Z H B I Z O Q M E P T
L H Z O A I M Q J W G P I E N O L C
B A A I H P D P E Z H D S K D R S F
K P O T P S Q I P A A A Z B V O Y I
N L O A O P Q Z S R H D Q V F C P S
A O M Z L W N E T P H S A A Q H M S
Q I S I E M O E R G D Y U T D R Q I
B D P L T R T E A R Z N Z M I O B O
M D E I A C T M S I V A Z T P M B N
E H R T R N E A N S V P N Z L O R S
N B M R I T Z J T I N S B Z O S T I
E A E E E F P Z V R F I J Z I O I S
G P L F E J O R H R S S B U D M T O
I S I S O T I M H F O A H Y G E E I
T G R B P U I L U A A F R A G B V E
T T H L W J Z Y G O T E P H E L G M
```

CLUES

1. Transmission of traits from parent to offspring
2. Division of chromosomes into two identical sets
3. Centriole plus astral rays
4. Produced by asexual reproduction
5. Union of unlike gametes
6. Male gamete
7. Pairing of homologous chromosomes in meiosis
8. Rod-shaped, carries genetic information
9. Between cell divisions
10. _____ fibers are associated with the chromosomes in mitosis and meiosis
11. Splitting or division
12. Fertilized egg

13. 1N
14. Group of four chromatids
15. Segment of DNA that controls a trait
16. Stage in which chromosomes contract and spindle fibers form
17. Stage of mitosis in which cytoplasm divides
18. Sex cell
19. Female gamete
20. Cell division resulting in haploid cells
21. 2N

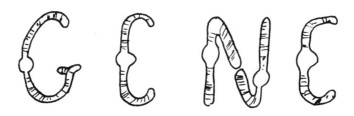

VOCABULARY

Basic

allele
cross-pollination
dihybrid cross
dominant
F_1 (first filial) generation
F_2 (second filial) generation
genetics
genotype
heterozygous
homozygous
law of independent assortment
law of segregation

Gregor Mendel
monohybrid cross
multiple alleles
P_1 (parental) generation
phenotype
pollination
principle of dominance and recessiveness
Punnett square
pure
recessive
self-pollination

DEFINITIONS

allele: one of a set of genes that control a particular trait

cross-pollination: the transfer of pollen from the anther of one plant to the stigma of another plant of the same species

dihybrid cross: cross that involves parents that are different in two pairs of traits

dominant: the one of a pair of traits that is expressed

F_1 (first filial) generation: the offspring of a cross between two parents that are pure of a certain trait

F_2 (second filial) generation: the offspring of the self-pollination of an F_1 generation

genetics: the science of heredity

genotype: the genetic makeup of an organism

heterozygous: describes an organism in which the pair of genes for a specific trait are not identical

homozygous: describes an organism in which the pair of genes for a specific trait are identical

law of independent assortment: states that genes are separated and distributed to gametes in a way that is completely independent of the other genes

law of segregation: states that a pair of genes is separated during the formation of gametes

Gregor Mendel: the father of genetics

monohybrid cross: cross that involves one pair of contrasting traits

multiple alleles: three or more alleles for the same trait

P₁ (parental) generation: parents are pure for a given trait

phenotype: the outward expression of the genotype

pollination: the transfer of pollen from an anther to a stigma

principle of dominance and recessiveness: states that one of the genes in a pair may mask the other gene, preventing the other from having an effect

Punnett square: a grid system used in computing the possible results of a genetic cross

pure: organism that always produces offspring with a certain trait

recessive: gene or characteristic that is masked in the presence of a dominant gene or allele

self-pollination: the transfer of pollen from anther to stigma in the same flower or flowers of the same plant

Name _____ Date _____

Class _____

PRINCIPLES OF HEREDITY:
BASIC TERMS CROSSWORD PUZZLES 1

ACROSS

1. Cross involving two pairs of traits
3. The law of _____ assortment results in much mixing of chromosomes in meiosis
4. One of a set of genes that controls a particular trait
5. Science of heredity
6. Gregor _____ was one of the first to carefully study heredity
8. Masked by dominant gene
10. Having unlike genes

DOWN

1. Masks recessive gene
2. Outward appearance
5. Genetic makeup
7. The law of _____ says that pairs of chromosomes or genes will separate during meiosis
9. Identical paired genes for a trait

PRINCIPLES OF HEREDITY:
BASIC TERMS CROSSWORD PUZZLE 2

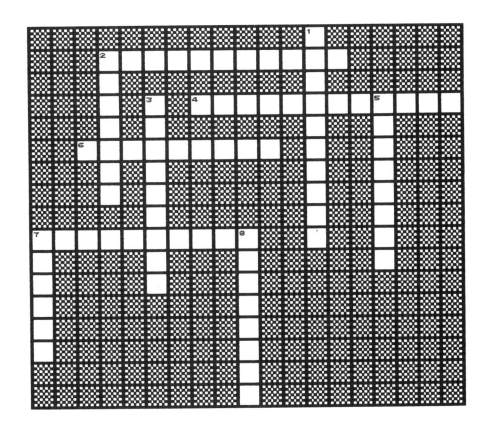

ACROSS

2. Transfer of pollen from anther to stigma
4. Having unlike alleles
6. Masked by a dominant gene
7. Cross in which there is one pair of contrasting traits

DOWN

1. Having the same allele
2. _____ squares help compute possible results of genetic crosses
3. Outward appearance
5. Genetic makeup
7. One of the first to carefully study heredity
8. Masks a recessive gene

PRINCIPLES OF HEREDITY:
VOCABULARY WORDSEARCH

The wordsearch below contains terms related to our study of the principles of heredity. The words can be found horizontally in either direction, vertically in either direction, and diagonally in either direction. Clues are given to help you find the words.

```
K R E C E S S I V E V W D M B S G S
F T U S J I A H K S R C J T Y Y E U
L U M W C L N H O B G G L G Q T N P
P G J L L I C D R M K N L V Y X O U
L O L E K Q T A E P O A B J T J T N
K C L G B T D E A P I Z N Q B V Y N
O E S L M V D R N L E J Y Q Q O P E
P D J M I C D V I E T N S G E D E T
X I U N V E F K M G I D B O Y P T
H R K N U F A C H P R M T E S U A X
Q B K B O R J T C D R W H I N Z S P
Q Y R B K D O M I N A N C E K T M N
L H V H G U K B I O E I W L B X Z H
C I A E D S M M J V N Y U M A S X T
F D B H I D O W A M L M C O P Z X D
L N M U P F D C C E P Y T O N E H P
U B E Q S U O G Y Z O R E T E H Z B
R X T D D I R B Y H O N O M Y F C R
I Y Q Y Z T N A N I M O D W E T O C
X L E D N E M N C C R X K G Y N Y J
```

ACROSS

1. Gregor _____ is referred to as the "father" of genetics
2. The first _____ generation is the offspring of a cross between parents that are pure for a given trait
3. The principle of _____ and recessiveness states that one gene may mask another allele
4. Outward expression or appearance of the genotype
5. One of a set of genes for a trait
6. A _____ cross involves parents that differ in two traits.
7. Study of heredity
8. The one of a pair of traits that is expressed
9. The law of _____ assortment states that genes are separated and distributed to gametes in a way that is independent of the other genes

DOWN

10. Having nonidentical genes for a trait
11. A _____ square can be used to compute the results of a cross
12. Transfer of pollen to stigma
13. Gene that is masked by a dominant gene or allele
14. Genetic makeup
15. Having identical genes for a trait
16. A _____ cross involves one pair of contrasting traits

VOCABULARY

Basic		*Advanced*
autosome	mapping	(all of the basic vocabulary)
chromsome map	mutagen	Thomas Hunt Morgan
chromosome mutation	mutant	operator
chromosome theory	mutation	operon
crossing over	nondisjunction	promoter
deletion	polyploidy	regulatory gene
Drosophila melanogaster	sex chromosomes	repressor
duplication	sex-linked gene	structural gene
gene linkage	sex-linked trait	
gene mutation	somatic mutation	
germ mutation	Walter Sutton	
incomplete dominance	X-chromosome	
inversion	Y-chromosome	

DEFINITIONS

autosome: a chromosome other than a sex chromosome

chromosome map: a diagram that shows the position of genes on chromosomes

chromosome mutation: a change that occurs in a chromosome that involves many genes

chromosome theory: states that chromosomes are the carriers of hereditary traits and that they carry the genes for many traits

crossing over: the exchange of segments of chromosomes that can occur when two chromosomes are in synapsis

deletion: chromosome mutation involving the loss of a piece of a chromosome

Drosophila melanogaster: species of small fruitfly used in genetic research

duplication: a chromosome mutation that occurs when an extra but identical piece of chromosome is added to the normal chromosome

gene linkage: genes on the same chromosome that stay together and cannot separate independently during meiosis

gene mutation: mutation that arises when the DNA code of a gene is changed

germ mutation: mutation that occurs in a reproductive cell

incomplete dominance: blend of two different traits

inversion: chromosome mutation that occurs when pieces of chromosomes break apart and the pieces reform the same chromosome in a different order

mapping: the process of locating specific genes on the chromosome

Thomas Hunt Morgan: geneticist who used *Drosophila* to discover sex-linked traits, gene linkage, and crossing over

mutagen: agent that causes mutation in genes or chromosomes

mutant: organism expressing a mutated, or changed, gene

mutation: any change in a gene or chromosome

nondisjunction: the failure of homologous pairs of chromosomes to separate during meiosis

operator: short segment of DNA in bacteria that controls a group of linked structural genes

operon: group of linked structural genes plus the controlling operator segment

polyploidy: condition in which cells contain more than the diploid number of chromosomes

promoter: short DNA segment in bacteria that binds with the amino acid sequence and turns off the structural gene

regulatory gene: gene that turns the structural gene on or off

repressor: in bacteria, the amino acid sequence of a protein that binds to the operator segment and turns off the structural gene

sex chromosomes: chromosomes of two kinds that determine the sex of the offspring

sex-linked gene: gene found on the X-chromosome

sex-linked trait: recessive trait carried on the X-chromosome

somatic mutation: mutation that occurs in a plant or animal body cell

structural gene: gene that codes for enzymes that control cell reactions and the production of body-building molecules

Walter Sutton: scientist who proposed that Mendel's "factors" for heredity were located on chromosomes

X-chromosome: in humans, the sex chromosome that is present singly in males and in pairs in females

Y-chromosome: sex chromosome found only in males

Name _____ Date _____

Class _____

MECHANISMS OF HEREDITY: BASIC TERMS CROSSWORD PUZZLE

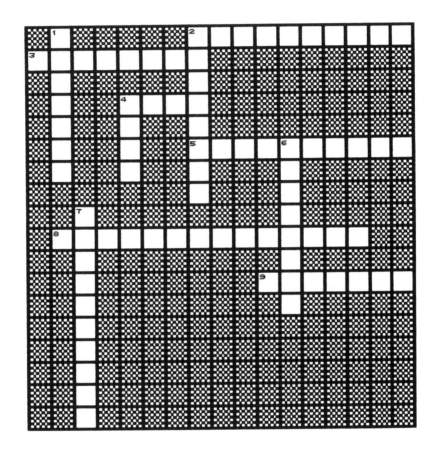

ACROSS

2. Fruitfly, often used to study heredity
3. Chromosome other than X or Y
4. _____ mutation results from a DNA change
5. Blend of two different traits
8. Failure of homologous chromosomes to separate during meiosis
9. _____ mutation occurs in the body but is not passed on to offspring

DOWN

1. Causes mutation
2. Leaving out or losing
4. _____ mutations occur in gametes and may be passed on to offspring
6. Sudden genetic change
7. Condition in which cells contain more than the diploid number of chromosomes

MECHANISMS OF HEREDITY:
ADVANCED TERMS CROSSWORD PUZZLE

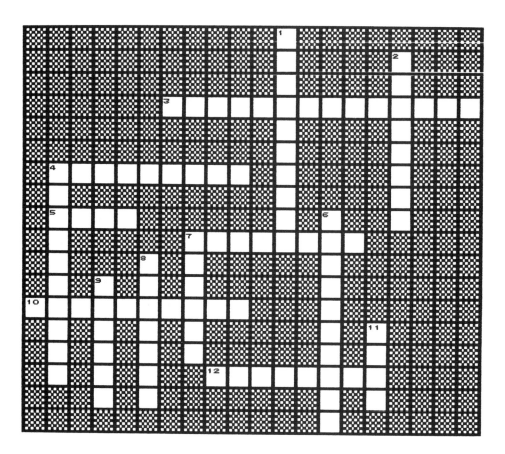

ACROSS

3. Failure of chromosomes to separate
4. Sequence of protein; turns off genes
5. _____ mutations occur in gametes
7. Segment of DNA that controls a group of linked genes
10. _____ genes code for enzymes that control reactions and material building
12. Sudden genetic change

DOWN

1. Fruitfly, often used in the study of genetics
2. Chromosome other than X and Y
4. _____ genes turn on and off the structural genes
6. More than diploid number of chromosomes
7. Group of linked structural genes plus their operator
8. Mutation-causing factor
9. Developed the chromosomal theory of heredity
11. Segment of DNA that controls trait

Name _____ Date _____

Class _____

MECHANISMS OF HEREDITY: VOCABULARY WORDSEARCH

The wordsearch below contains terms related to our study of the mechanisms of heredity. The words can be found horizontally in either direction, vertically in either direction, and diagonally in either direction. Clues are given to help you find the words.

```
J Y O K L M G N I P P A M Z M Y H X
G T J Y R P L A T N A T U M Q O S L
Y M P W H N E G A T U M Q U O E H Z
N P O L Y P L O I D Y S Y W M H B W
O K D P D E K N I L X E S O O S F E
N E R F F R P A Z L I T S M V P M G
D N E O S O D O P K C O H F P U L E
I E G E Y X S N I L M S Q M B V G Q
S G U D M Q C D R O S O P H I L A J
J S L D I F Y B R M U T A T I O N E
U E A W X I N H D S N H B Z C C T T
N M T R H U C R J D C S N Z U U A E
C O O E O X E M O S O M O R H C Y L
T S R U E Z G X H S W J E T N V S P
I O Y S B I Q L U E S Y N F S T T M
O T G E M W X N W W E E E N T F C O
N U E V A S G B O C O G R R P S O C
T A N O P E R O N E K Y N P G S L N
E X E A W H K F A O K X Q B E L K I
P S O M A T I C D X P D B Z C R S M
```

CLUES

1. Chromosomes that determine the sex of an offspring (two words)
2. Genes found on the X-chromosome (two words)
3. _____ dominance is a blend of two different traits
4. A _____ mutation occurs in a body cell
5. _____ turns the structural gene on and off (two words)
6. An agent that causes a mutation
7. Chromosome other than a sex chromosome
8. _____ *melanogaster* is a species of fruitfly
9. A _____ mutation arises from a DNA change
10. Turns off a structural gene
11. Change in a gene or chomosome
12. Failure of homologous pairs of chromosomes to separate
13. A sex chromosome that is found only in males
14. Process of charting positions of genes on chromosomes
15. Condition in which cells contain more than 2N chromosomes
16. Group of linked structural genes plus controlling operator segment
17. Organism showing a mutation

VOCABULARY

Basic

acquired trait	homologous structures
adaptation	industrial melanism
analogous structures	Jean Baptiste Lamarck
artificial selection	Mesozoic era
Cenozoic era	migration
comparative anatomy	Stanley Miller
comparative biochemistry	natural selection
comparative embryology	Alexander Oparin
Charles Darwin	Paleozoic era
epoch	period
era	Precambrian era
eukaryotic cell	prokaryotic cell
evolution	race circle
fossil	random mating
gene flow	reproductive isolation
genetic drift	sedimentary rock
genetic equilibrium	speciation
geologic time	vestigial organ

Advanced

(all of the basic vocabulary)
allele frequency
Hardy-Weinberg Principle

DEFINITIONS

acquired trait: body change caused by an organism's adapting to its environment

adaptation: any genetically controlled characteristic that aids an organism to survive and reproduce in its environment

allele frequency: term that indicates how often a certain allele occurs in a given population

analogous structures: parts of different organisms that are similar in function but of different evolutionary origin

artificial selection: the process used by plant and animal breeders to produce desired genetic traits in offspring

Cenozoic era: the age of mammals, from 70 million years ago to the present

comparative anatomy: the study and comparison of the body parts of different species

comparative biochemistry: the study and comparison of the chemical makeup of the DNA and proteins of different species

comparative embryology: the study and comparison of the embryonic development of different species

Charles Darwin: scientist who proposed the theory of evolution by natural selection

epoch: division of geologic time within a period

era: the largest division of geologic time

eukaryotic cell: a cell containing a distinct, membrane-bound nucleus

evolution: change in the genetic makeup of a population over time

fossil: the remains of a dead organism

gene flow: the movement of genes in or out of a population

genetic drift: change in the gene pool as the result of chance, not of selection, migration, or mutation

genetic equilibrium: condition in which the gene pool does not change or evolve

geologic time: time scale that shows the history of the Earth characterized by certain changes that took place

Hardy-Weinberg Principle: principle that states the conditions under which a gene pool does not change or evolve

homologous structure: body parts in different organisms that are different in function but similar in structure and evolutionary origin

industrial melanism: rapid genetic shift from light coloration to dark coloration in response to industrial pollution

Jean Baptiste Lamarck: scientist who explained evolution through the inheritance of acquired characteristics

Mesozoic era: the age of dinosaurs, from 225 million to 70 million years ago

migration: the movement of an organism from one location to another

Stanley Miller: the first person to test Oparin's theory of the origin of life

natural selection: the process that leads to an increase in the frequency of some genes and the decrease in the frequency of others

Alexander Oparin: Russian scientist who first proposed a hypothesis to explain the origin of life, namely, that simple gases in the Earth's early atmosphere could have been converted to complex organic compounds

Paleozoic era: the age of amphibians, from 600 million to 225 million years ago

period: geologic time division of an era

Precambrian era: the oldest geologic era, from 4,500 million to 600 million years ago

prokaryotic cell: a type of cell that lacks a membrane-bound nucleus

race circle: a group of subspecies within a certain geographic area

random mating: mating in which there is no preference shown for a particular phenotype

reproductive isolation: occurs when members of two groups of the same species cannot mate with each other, usually caused by a geographic barrier

sedimentary rock: rock formed as soil particles are laid down, forming new layers over old ones

speciation: the process of formation of new species

vestigial organ: structure in animals that appears to have no function

ORIGINS OF LIFE AND THEORIES OF EVOLUTION:
BASIC TERMS CROSSWORD PUZZLE

ACROSS

1. Cell that lacks nuclear membrane
5. Industrial _____ resulted in dark-colored organisms
7. Characteristic that enables the individual to survive better
8. Developed theory of evolution by natural selection
9. _____ era, age of amphibians
10. Seems to have no function
11. Cells with nuclear membrane

DOWN

2. Developed theory of the origin of life on Earth
3. Oldest geologic era
4. _____ structures have different functions but are similar in structure and origins
6. Formation of new species
7. _____ structures have similar functions but different origins

Name _____ Date _____

Class _____

ORIGINS OF LIFE AND THEORIES OF EVOLUTION: ADVANCED TERMS CROSSWORD PUZZLE

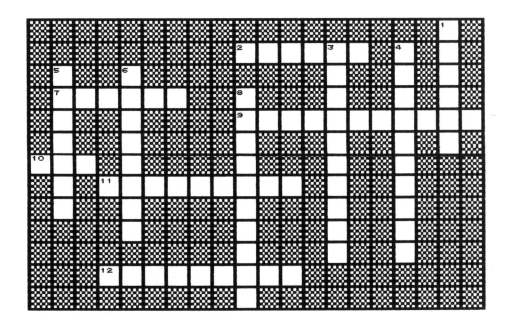

ACROSS

2. Tested Oparin's theory
7. _____ frequency indicates how often a certain allele occurs in a given population
9. Cell without nuclear membrane
10. Largest division of geologic time
11. Seemingly without function
12. _____ structures have similar functions but different origins

DOWN

1. Developed theory of evolution by natural selection
3. Cells with nuclear membrane
4. _____ structures have different functions but are similar in origin
5. Developed theory of evolution by inheritance of acquired characteristics
6. The Hardy-_____ Principle states conditions that inhibit evolution
8. Formation of new species

ORIGINS OF LIFE AND THEORIES OF EVOLUTION: VOCABULARY WORDSEARCH

The wordsearch below contains terms related to our study of the origins of life and the theories of evolution. The words can be found horizontally in either direction, vertically in either direction, and diagonally in either direction. Clues are given to help you find the words.

```
O Y G R Z A K T U Y B N I W R A D M
O P M E S L A N W Z U I O U C H V A
A L J D V L D O Z R K J Y Y B U C L
D R A H H E A I A N A L O G O U S L
U W W G O L P T L N Y H N Y X C B U
T L G C M E T U Y N H S S J G I P P
I P E E O Z A L E O A P W F L P X X
T R O R L U T O M I R E R P I I L Q
J E L Z O O I V W T D C I A U R A K
B C O K G H O E O A Y I A L V S I Z
C A G U O T N N D R W A R E J A G A
I M I N U Z N B X G E T Q O O U I M
O B C I S P A C R I I I D Z R T T I
Z R T R Q Q K E E M N O J O K Z S V
O I I A U D D F R R B N L I O B E U
N A M P W D N F E P E H N C E I V E
E N E O I D Y I M E R T Z L E F R M
C U B K C R A M A L G O X K L A L X
D Q N M T L L K D D S S T T J Z H T
X R S Q U Y X Q E V X S I K P L S V
```

© 1991 by Center for Applied Research in Education

CLUES

1. Proposed a theory to explain the origin of life on Earth
2. The _____ era is the "age of mammals"
3. Oldest geologic era
4. Structures that are similar in function but not in origin
5. Charles _____ proposed a theory of evolution by natural selection
6. _____ frequency indicates how often a certain allele occurs in a population
7. Time scale showing history of the Earth (two words)
8. Formation of new species
9. Change in a population's genetic makeup with time
10. _____ organs have no apparent function
11. A genetically controlled characteristic that aids survival
12. Movement from one place to another
13. Largest division of geologic time
14. Age of amphibians
15. Different in function but similar in origin
16. He proposed theory of inheritance of acquired characteristics
17. Principle that states the conditions under which a gene pool does not change (two words)

VOCABULARY

Basic		*Advanced*
amoeba	gullet	(all of the basic vocabulary)
ameboid movement	macronucleus	acrasin
anal pore	micronucleus	Acrasiomycota
binary fission	mouth pore	cellular slime mold
cilia	oral groove	Ciliophora
ciliate	*Paramecium*	Mastigophora
contractile vacuole	pellicle	myxameba
cyst	Protista	Myxomycophyta (Myxomycota)
ectoplasm	pseudopodium	plasmodium
endoplasm	sporangium	pseudoplasmodium
flagella	vacuole	Sarcodina
flagellate	zooflagellate	slime mold
food vacuole		Sporozoa
		trichocyst
		water mold
		zoospore

DEFINITIONS

acrasin: a substance secreted by myxameba that causes them to swarm together to form a clump of cells

Acrasiomycota: a phylum that includes cellular slime molds

amoeba: a protist characterized by its distinctive "shapeless" body structure

ameboid movement: the creeping motion of a cell caused by the pressure of cytoplasm against the cell membrane

anal pore: an opening in the pellicle of a *Paramecium* through which undigested particles are eliminated

binary fission: asexual reproduction in which one cell divides into two cells of equal size

cilia: tiny hairlike projections found on the surface of some cells

ciliate: member of a group of protists that move by means of cilia

Ciliophora: phylum of protists that move by means of cilia

contractile vacuole: organelle found in single-celled organisms that helps to maintain water balance in the cell

cyst: hard-walled structure encasing the resting stage of an organism

ectoplasm: the clear watery cytoplasm found immediately inside the cell membrane

endoplasm: the dense grainy cytoplasm found in the interior of some cells

flagella: thin whip-like structures found on the outside of some cells

flagellate: member of a group of protists that move by means of a flagellum

food vacuole: vacuole that forms around food

gullet: funnel-like structure in *Paramecium* that extends from the mouth pore to the endoplasm

macronucleus: one of two nuclei in *Paramecium* that controls respiration, protein synthesis, and digestion

Mastigophora: phylum that includes organisms that move by means of flagella

micronucleus: the smaller of two nuclei found in *Paramecium* that functions only during reproduction

mouth pore: opening at one end of the oral groove of a *Paramecium*

myxameba: the single-celled stage of a cellular slime mold

Myxomycophyta (Myxomycota): the phylum that includes the plasmodial slime molds

oral groove: cavity along one side of a *Paramecium*

Paramecium: common ciliated protist

pellicle: a thick outer membrane surrounding the cell of the *Paramecium* and other organisms

plasmodium: the body of a slime mold containing many nuclei and no cell walls

Protista: the kingdom in which the single-celled organisms are found

pseudoplasmodium: mass of single-celled myxamebae

pseudopodium: the "false foot" of an amoeba

Sarcodina: the phylum that includes *Amoeba*

slime mold: fungus-like protist that includes an amoeba-like feeding stage and a stationary reproductive stage

sporangium: reproductive body of a fungus-like protist

trichocyst: thread-like organelle inside the pellicle of a *Paramecium* that serves as a defense response

vacuole: fluid-filled space inside a cell surrounded by a membrane

water mold: fungus-like aquatic protist that is filamentous and multinucleate

zooflagellate: member of a group of protists that moves by means of one or more flagella

zoospore: flagellated cell that leaves the parent cell and develops into a new organism

Name _____ Date _____

Class _____

PROTISTS:
BASIC TERMS CROSSWORD PUZZLE

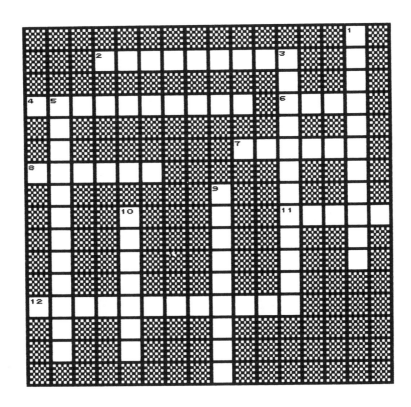

ACROSS

2. Dense, grainy cytoplasm
4. Reproductive body of fungus-like protist
6. Hard-walled structure
7. Protist with "shapeless" body
8. Funnel-like passage to mouth
11. Tiny hair-like projections
12. Smaller of two nuclei

DOWN

1. The _____ vacuole helps to maintain water balance
3. Larger of two nuclei in some organisms
5. "False foot"
9. Clear, watery cytoplasm
10. Food _____ forms around food

Name _____   Date _____

                                          Class _____

PROTISTS:
ADVANCED TERMS CROSSWORD PUZZLE

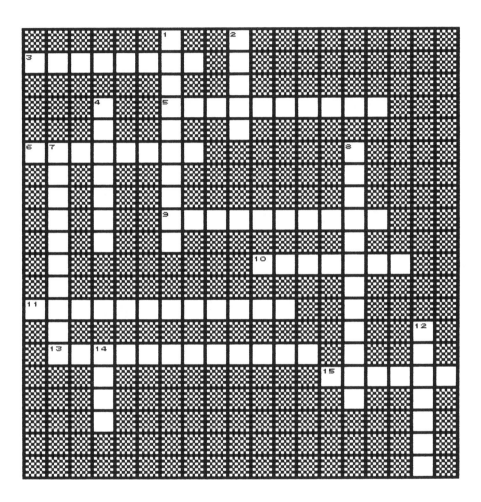

ACROSS

  3. Flagellated cell that leaves parent cell and develops into a new organism
  5. Phylum of protists that moves by cilia
  6. Phylum of nonmotile parasitic protists
  9. Reproductive body of fungus-like protist
10. Contractile _____ helps regulate water balance
11. Smaller of two nuclei in *Paramecium*
13. Larger of two nuclei in *Paramecium*
15. Funnel-like food passage

DOWN

  1. Thread-like organelle; defense in *Paramecium*
  2. Tiny hair-like structures
  4. Secreted by myxameba, causing them to clump
  7. Body of slime mold
  8. "False foot"
12. _____ movement is a creeping motion caused by pressure of cytoplasm on cell membrane
14. Hard-walled structure encasing resting stage

PROTISTS:
VOCABULARY WORDSEARCH

The wordsearch below contains terms related to our study of protists. The words can be found horizontally in either direction, vertically in either direction, and diagonally in either direction. Clues are given to help you find the words.

```
M P T Z V O K S V U X C K E T K N V
A V R R C R B Z M A O Z O R O P S E
H M M Y X O M Y C O P H Y T A M Q Q
D O O A E N N V A M R E E V W A N E
N X K E C W P T K T F Z J A P C Q L
W M F E B N S P R A S C S C B R F C
K S O S Y A E P E A C Y Z U D O M I
F A A E J J U A L B C L C O V N K L
S L X M E U D N C A X T X L G U F L
B P H V T I O I O I S Q I E I C L E
A O T Z A B P D N I N M M L Q L C P
V T R U L I O O P Z S U O J E E H E
M C D U L N D C R V U S U D F U X E
E E N D E A I R O Z A V I L I S M A
D V T F G R U A T Q E L P Z Q U Z I
Y X X O A Y M S I B W E O S L L M L
X C J T L N W W S K R V F J T R E I
U T G W F X C W T L M E T A I L I C
V Z Z G K G H O A T E M V G P V X O
P A U W T K Q M U I C E M A R A P Y
```

CLUES

1. Kingdom of protists
2. Phylum of nonmotile spore-forming parasites
3. A _____ vacuole helps maintain water balance
4. Hard-walled structure
5. Tiny hair-like projections
6. Larger of two nuclei; controls respiration, protein synthesis, and digestion in *Paramecium*
7. Phylum of amoeba
8. "False foot"
9. Fluid-filled cavity or sac
10. _____ fission results in two equal-sized cells
11. Thick outer membrane of *Paramecium*
12. Body of slime mold, containing many nuclei
13. Having cilia
14. Protist with "shapeless" body structure
15. Clear, watery cytoplasm just inside cell membrane
16. Common ciliated protist
17. Member of a group of protists that move by flagella
18. Phylum of slime molds

VOCABULARY

Basic		*Advanced*
alga	gametophyte	(all of the basic vocabulary)
alternation of	holdfast	alginate
generation	phytoplankton	*Chlamydomonas*
antheridium	plankton	Chlorophyta (green algae)
archegonium	red tide	Chrysophyta (golden-brown algae)
blade	reservoir	Euglenophyta (euglenoids)
bioluminescence	*Spirogyra*	fucoxanthin
colonial	sporophyte	Phaeophyta (brown algae)
diatom	stipe	phycobilin
diatomaceous earth	thallus	pyrenoid
epiphyte	unicellular	Pyrrophyta (dinoflagellates)
Euglena	*Volvox*	Rhodophyta (red algae)
eyespot	zygospore	
filamentous		
flagella		

DEFINITIONS

alga: a protist that is usually aquatic, contains chlorophyll, and carries out photo-synthesis

alginate: a chemical found in brown algae, used as a thickening agent

alternation of generation: life cycle in organisms in which a haploid stage produces gametes that join to form a zygote, which germinates to form a diploid phase; meiosis produces spores that produce a new haploid generation

antheridium: male gamete-producing structure found in spore plants, some algae, and some fungi

archegonium: female gamete-producing structure found in spore plants, some algae, and some fungi

blade: the broad, flat, leaf-like part of a thallus of brown algae

bioluminescence: the ability of some autotrophic protists to glow in the dark when disturbed

Chlamydomonas: common unicellular green alga

Chlorophyta: phylum of algae that includes both single-celled and multi-cellular green algae

Chrysophyta: phylum of algae that includes the single-celled golden-brown algae

colonial: single-celled algae living together in a group

diatom: golden-brown alga with cell walls composed of silica

diatomaceous earth: deposits of diatom remains

epiphyte: organism that grows on other organisms without harming them

Euglena: typical protist that has chlorophyll and carries out photosynthesis, but can also absorb some nutrients from the environment

Euglenophyta: a phylum of protists that includes the *Euglena*

eyespot: sensory structure in organisms like *Euglena* that is sensitive to light

filamentous: having a long, thread-like structure

flagella: thin, whip-like structures found on the outside of some cells

fucoxanthin: brown pigment found in members of the phylum Phaeophyta

gametophyte: the haploid stage that produces gametes in alternation of generation

holdfast: structure containing special cells that anchor the base of an alga to a substrate

Phaeophyta: phylum of algae that includes the brown algae

phycobilin: red pigment found in the red algae

phytoplankton: floating microscopic algae found in lakes and oceans

plankton: very small plants and animals that float in surface waters of lakes and oceans

pyrenoid: spherical area of a chloroplast in which sugar is converted to starch and stored

Pyrrophyta: phylum of algae that includes the dinoflagellates

red tide: dense population of some algae that produces nerve toxins dangerous to humans who eat contaminated shellfish

reservoir: structure in *Euglena* that contains the cell's excess water

Rhodophyta: a phylum of algae that includes the red algae

Spirogyra: common, filamentous green alga

sporophyte: diploid structure that undergoes meiosis and produces haploid spores

stipe: stem-like structure of a kelp thallus

thallus: the body of a multicellular alga such as kelp

unicellular: algae that live in a single-celled state

Volvox: colonial green alga consisting of thousands of cells connected by strands of cytoplasm

zygospore: zygote with a thick protective wall

AUTOTROPHIC PROTISTS:
BASIC TERMS CROSSWORD PUZZLE

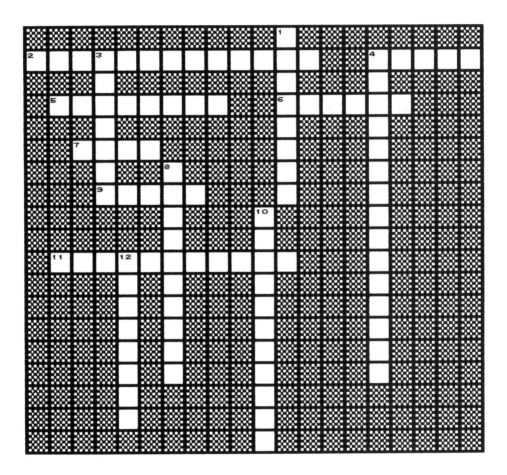

ACROSS

2. Microscopic floating algae
4. Leaf-like part of thallus
5. Floating organisms
6. Golden-brown algae, silica cell walls
7. Photosynthetic protist
9. Stem-like part of thallus
11. Haploid stage in alternation of generations

DOWN

1. Anchoring part of alga
3. Body of multicellular alga such as kelp
4. Glowing in dark by organism
8. Diploid stage
10. Male gamete-producing structure in algae
12. Organism that grows on other organisms without harming them

Name _____   Date _____

Class _____

AUTOTROPHIC PROTISTS:
ADVANCED TERMS CROSSWORD PUZZLE

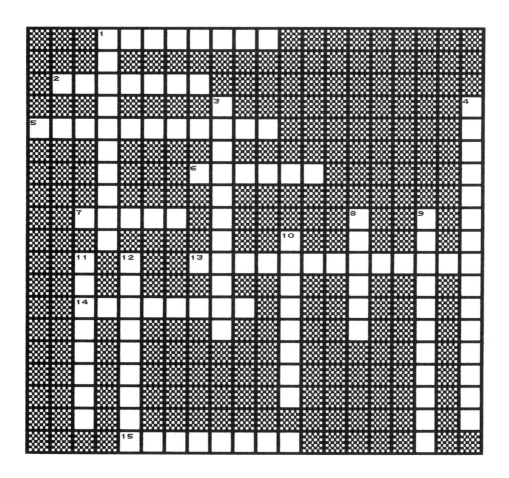

ACROSS

1. Part of chloroplast in which sugar is converted to starch
2. Body of kelp
5. Brown pigment
6. Colonial green alga
7. Stem-like part of kelp
13. Floating microscopic algae
14. Chemical found in brown algae, used as thickening agent
15. Grows on other organism

DOWN

1. Brown algae
3. Green algae
4. Ability to glow in dark
8. Golden-brown algae, cell wall made of silica
9. Produces male gametes in algae
10. Kelp anchor
11. Floating organism
12. Zygote with thick protective wall

AUTOTROPHIC PROTISTS:
VOCABULARY WORDSEARCH

The wordsearch below contains terms related to our study of autotrophic protists. The words can be found horizontally in either direction, vertically in either direction, and diagonally in either direction. Clues are given to help you find the words.

```
I F R A M O A R Y G O R I P S R Z K
N B U O F Q L O B A L G I N A T E J
B V S S I H H V V D P D Y G M N X
H F K I R A T Y H P O E A H P J Q S
A W Q P A Z C O N O T K N A L P J H
O V Y T Q F F A T Y H P O R O L H C
D O P W M U I D I R E H T N A C X H
D N B M R G E T Y H P O R O P S D T
P O O U L C N M O T A I D G R M C G
U T B I O L U M I N E S C E N C E W
T K S N I X N H D I G V V J M K F J
F N F O J K C E T Y H P O T E M A G
E A M G I Q S A L G A E B K N C V V
H L O E R F C V V I E N B E K H T F
L P F H U U X B F B I T N A V O W Q
P O C C K S U O T N E M A L I F L N
H T L R T R R A L U L L E C I N U A
G Y R A T H I P O T D G A P S T I T
S H W Z B B N E T Y H P I P E S E A
L P L R U Y A L M H I K U U W A Y W
```

CLUES

1. Photosynthetic, usually aquatic, protist
2. One-celled
3. Phylum of brown algae
4. Haploid stage; produces gametes in alternation of generation
5. Produces male gametes in spore plants
6. Glowing in dark by organisms
7. Plant plankton
8. Long and thread-like
9. A common filamentous green alga
10. Diploid stage; produces spores in alternation of generation
11. Organism that grows on another without harming it
12. Floating organisms
13. Phylum of green algae
14. Thickening agent derived from algae
15. Female gamete-producing structure in spore plants
16. Golden-brown algae with silica cell walls

11: FUNGI

VOCABULARY

Basic		*Advanced*
ascospore	mold	(all of the basic vocabulary)
ascus	mushroom	apothecium
basidiospore	mycelium	Ascomycota
basidium	mychorrhiza	Basidiomycota
decomposer	parasite	Fungi imperfecti
Fungi	rhizoid	gametangium
fungus	sorus	heterokaryon
gills	sporangium	homokaryon
hypha	stipe	ostiole
lichen	yeast	perithecium
mildew	zygospore	primary mycelium
		secondary mycelium
		sporangiophore
		Zygomycota

DEFINITIONS

apothcium: saucer-shaped fruiting body found in certain Ascomycetes, such as the morel or the truffle

Asocomycota: phylum of fungi that includes yeasts, molds, and mildews

ascospore: haploid spore produced by the fusion of male and female nuclei in the ascus

ascus: sac found on the tip of specialized hyphae in which sexual spores are produced

Basidiomycota: phylum of fungi that includes mushrooms, rusts, and smuts

basidiospore: haploid spore produced by the basidium

basidium: club-like structure that produces basidiospores

decomposer: organism that lives on dead or decaying organic matter

Fungi: the kingdom of organisms that lack chlorophyll and must absorb nutrients through cell walls made of chitin

fungus: any member of the kingdom Fungi

Fungi imperfecti: phylum of fungi that lacks a means of sexual reproduction

gametangium: structure that produces gametes

gills: structures on the underside of a mushroom cap on which basidia form

heterokaryon: fungus that has two or more genetically different nuclei in its mycelium

homokaryon: fungus that has one type of nucleus in its hyphae

hypha: tube-like structure that makes up the body of a fungus

lichen: association of a fungus and an alga

mildew: fungal disease of plants

mold: superficial growth of a fungus mycelium

mushroom: edible fruiting body of a fungus

mycelium: mass of interwoven hyphae

mychorrhiza: association of a fungus and the roots of a plant

ostiole: narrow opening in the perithecium through which spores are released

parasite: organism that lives in or on another organism

perithecium: flask-like fruiting body of some ascomycetes

primary mycelium: hypha with one nucleus in each cell

rhizoid: short hypha that anchors a fungus to a substrate

secondary mycelium: fused hyphae of two different mating strains with two nuclei in each cell

sorus: the asexual fruiting body found in rusts and smuts

sporangiophore: upright stalk in molds that produces spore cases at its tip

sporangium: spore case found in molds

stipe: stalk for the mushroom cap

yeast: fungus that does not have hyphae or fruiting bodies and reproduces asexually by budding

Zygomycota: phylum of fungi that includes the black bread mold *Rhizopus*

zygospore: zygote with a thick protective wall

Name _____ Date _____

Class _____

FUNGI:
BASIC TERMS CROSSWORD PUZZLE

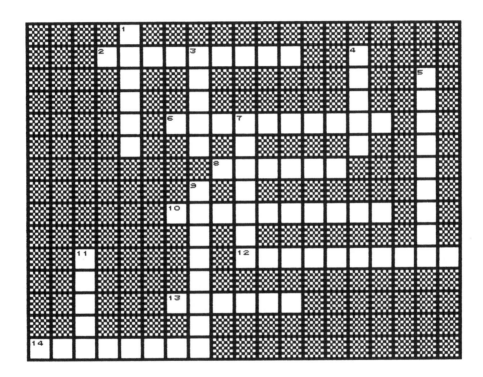

ACROSS

2. Zygote with thick protective wall
6. Spore case
8. Plant fungal disease
10. Association of fungus and roots
12. Lives on dead organisms
13. Fungus and alga living together
14. Club-like structure that produces basidiospores

DOWN

1. Make up body of fungus
3. Stalk of mushroom
4. Asexual fruiting body of rust
5. Spore produced in ascus
7. Short anchoring hypha
9. Mass of interwoven hyphae
11. On underside of mushroom cap; location of basidia

FUNGI:
ADVANCED TERMS CROSSWORD PUZZLE

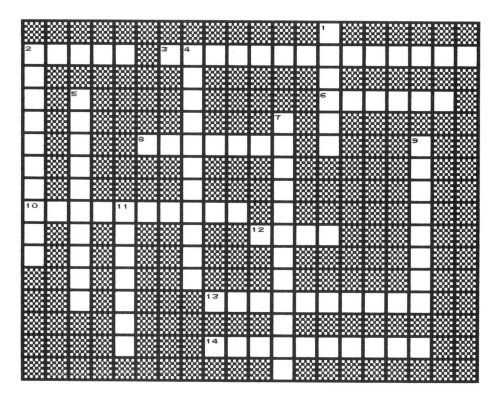

ACROSS

2. Sac in which sexual spores are produced
3. Upright stalk in molds; produces spore cases
6. Makes up body of fungus
8. Opening through which spores are released
10. Fungi _____ do not reproduce sexually
12. Superficial mycelium growth
13. Spore case
14. Fungus living with and on roots

DOWN

1. Fungus living with alga
2. Saucer-shaped fruiting body
4. Flask-like fruiting body
5. Lives on dead matter
7. Has two or more genetically different nuclei in mycelium
9. Mass of hyphae
11. Short anchoring hypha

Name _____   Date _____

Class _____

FUNGI:
VOCABULARY WORDSEARCH

The wordsearch below contains terms related to our study of fungi. The words can be found horizontally in either direction, vertically in either direction, and diagonally in either direction. Clues are given to help you find the words.

```
J B M H Q M F T A O W Y A G J C L O
C K U C F J L M U I L E C Y M H W F
E F S W X K M A M E V I Q E E F K T
L T H Y E R O P S O G Y Z T G S U U
N U R D E U M N M V M V E F X A J H
I G O C P A R T D O D R D V J E M R
V Z O S I C S A L R O R E H H L U F
E R M V T A K D U K Z E T Y C L I W
L E H I S V M T A C Z X I P E I D J
K S A N Q G Q R S Z N K S H P G I I
N O D M E O Y R C F F B A A I T S D
E P G E B O M H O A I K R E F M A K
H M V A N P B I M F I H A R A M B O
C O P K C T S Z Y U J G P Y I X V B
I C L I S C V O C N F Y N L Y J Q Q
L E U A U P R I O G Z I D A B V R H
Q D E F R U B D T I P E X U R W D V
D Y H W O G Q N A O W X N A V O Q U
T D Z Z U X O S T I O L E D F Z P E
I A H M Y C O R R H I Z A V X A A S
```

CLUES

1. Kingdom of organisms that lack chlorophyll and absorb nutrients through chitinous cell walls
2. Tube-like structures; form the body of fungus
3. Phylum of yeasts and molds
4. Zygote with thick protective wall
5. Superficial mycelium growth
6. Edible fruiting body of fungus
7. Stalk of mushroom
8. Organism that lives on another at the host's expense
9. Mass of interwoven hyphae
10. Short, anchoring hypha
11. Association of fungus with an alga for mutual benefit
12. Plural of ascus
13. Fungus without hyphae; reproduces asexually by budding
14. Structure on underside of mushroom cap
15. Lives on dead or dying organisms
16. Fungus with two or more genetically different nuclei
17. Plural of sporangium
18. Fungal disease of plants
19. Opening through which spores are released
20. Club-like structure; produces basidiospores
21. Association of a fungus and a plant's roots

VOCABULARY

Basic		*Advanced*
angiosperm	internode	(all of the basic vocabulary)
apical meristem	meristem	collenchyma
cambium	monocot	companion cell
conifer	node	cycad
cork	ovulate cone	ginkgo
cotyledon	pollen tube	megaspore
dicot	root system	microspore
embryo sac	seed	parenchyma
epidermis	shoot system	sieve element
gymnosperm	staminate cone	sieve plate
		sieve tube
		tracheid
		vessel element

DEFINITIONS

angiosperm: flowering plant

apical meristem: actively dividing tissue found at the tips of stems, branches, and roots of plants

cambium: ring of meristem cells found between xylem and phloem of vascular bundles; responsible for the increase in thickness of shoots and roots

collenchyma: plant tissue composed of long, thin, easily stretched cells that strengthen the stem and leaf stalks

companion cell: small cell that lies alongside the sieve tube in phloem tissue and appears to control the activities of the sieve elements

conifer: tree that bears its sex organs in cones

cork: the outermost cell of woody plants which contain substances that prevent water loss

cotyledon: the seed leaf of a plant

cycad: gymnosperm that resembles a small palm tree

dicot: plant with two seed leaves

embryo sac: the female gametophyte in angiosperms

epidermis: the outer layer of cells in roots, stems, leaves, flowers, and seeds

ginkgo: gymnosperm, "living fossil" that loses its leaves in the fall

gymnosperm: seed plant that bears exposed seeds

internode: stem region between two nodes

megaspore: large spore that develops into a female gametophyte

meristem: the growing point of a plant

microspore: small spore that develops into a male gametophyte

monocot: plant with one seed leaf

node: place on a stem where a leaf is attached

ovulate cone: female reproductive organ of a gymnosperm

parenchyma: plant tissues made up of large, loosely packed cells that are active in photosynthesis, food storage, conduction, and secretion

pollen tube: male gametophyte in seed plants

root system: structure that anchors plants to the ground, absorbs water and minerals from the soil, and conducts them to the stem

seed: structure that contains the embryo or young sporophyte plant and a supply of food to help the embryo grow

shoot system: system of structures in plants that is usually above ground and consists of stems, leaves, and flowers

sieve element: phloem tissue that transports food

sieve plate: perforated wall between two sieve elements

sieve tube: conducting tube of the phloem

staminate cone: male reproductive organ of gymnosperms

tracheid: long, thick-walled, water-conducting cell that strengthens woody tissue

vessel element: thin-walled conducting tissue in angiosperms

Name _____ Date _____

Class _____

SEED PLANTS:
BASIC TERMS CROSSWORD PUZZLE

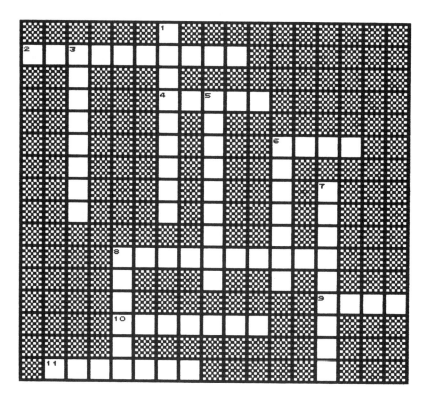

ACROSS

2. Seed plant with exposed seeds
4. Plant with two seed leaves
6. Outermost cells of woody plants
8. Flowering plant
9. Site of leaf attachment
10. Growth layer between xylem and phloem
11. Female cone

DOWN

1. Outer cell layer
3. Growth area
5. Seed leaf
6. Cone-bearing plant
7. Male cone
8. Tip

Name _____   Date _____

Class _____

SEED PLANTS:
ADVANCED TERMS CROSSWORD PUZZLE

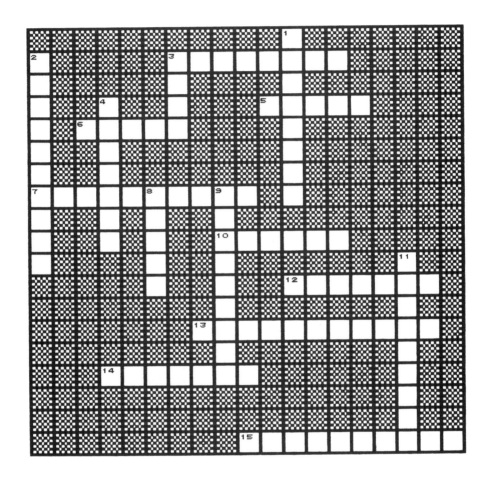

ACROSS

3. Thick-walled water-conducting cell
5. Plant with two seed leaves
6. Sieve _____ is a perforated wall
7. Large, loosely packed cells
10. Gymnosperm "living fossil" that loses leaves in fall
12. Female cone
13. Long, thin, easily stretched cells; add strength to leaves
14. Cone-bearing plant
15. Flowering plant

DOWN

1. Growth area
2. Small spore that develops into male gametophyte
3. Sieve _____ is conducting tube of phloem
4. Sieve _____ is food-transporting phloem tissue
8. Palm-like gymnosperm
9. Large spore that develops into female gametophyte
11. Male cone

SEED PLANTS:
VOCABULARY WORDSEARCH

The wordsearch below contains terms related to our study of seed plants. The words can be found horizontally in either direction, vertically in either direction, and diagonally in either direction. Clues are given to help you find the words.

```
H A M Y H C N E R A P L P D V B V F
J M O U O O I Z M E T S I R E M G I
T E A N D D A C Y C I S Z C F U Y E
L N M E J E Y V Y N E L L O P G M M
N O E I H D I E H C A R T V G M N B
Q D A B C X F T X P N M V O V I O R
D E I C E P B R R M L T E E H C S Y
I L S T A M I N A T E I E M K R P O
C Y P K E N K B K H I B D J K O E C
O T Z H A N G I O S P E R M X S R T
T O Z C A S L E A Z A B J S Y P M T
R C M N B B L S Q L K R B C Q O Q Y
X X W T P V N Q E O B R D X F R I D
S I M O W S I M R E D I P E Z E D O
H S Z N M O N O C O T X Q I L B Q B
I I Q X U R E F I N O C M Q Z L L M
W Z I D G L I X M O G K N I G O G Z
O M O M R A G B W Z L R A L V J A C
J P V N A C R I W N W B L R C H O C
A V K B M G Q C A M B I U M K K O S
```

CLUES

1. Seed plant with exposed seeds
2. The _____ tube produces pollen
3. A deciduous "living fossil"
4. Plant with one seed leaf
5. Tissue made up of large loosely packed cells; active in photosynthesis, food storage, conduction, and secretion
6. Meristem between xylem and phloem
7. Flowering plant
8. The _____ sac produces the female gametes in angiosperms
9. Cone-bearing plant
10. Plant with two seed leaves
11. Site of leaf attachment
12. Outer layer of cells
13. Small spore that develops into male gametophyte
14. Gymnosperm that resembles a small palm tree
15. The _____ cone is the male reproductive organ of gymnosperms
16. Seed leaf
17. Long thick-walled conducting cell
18. Growing part of plant

VOCABULARY

Basic		*Advanced*
active transport	phloem	(all of the basic vocabulary)
adventitious root	primary growth	aerial root
cork cambium	primary root	air root
cortex	primary tissue	auxin
dormant	root	capillary water
endodermis	root cap	gibberellin
epidermis	root hair	pericycle
fibrous root	secondary growth	prop root
geotropism	secondary tissue	region of cell division
herbaceous plant	taproot	region of elongation
humus	tropism	region of maturation
loam	vascular cambium	transpiration
macronutrient	woody plant	
micronutrient	xylem	
osmosis		

DEFINITIONS

active transport: the passage of substances across a semipermeable membrane that requires the use of energy

adventitious root: root that develops from the node of a stem or from a leaf

aerial root: root of a plant that lives off the ground and absorbs water from the air

air root: special root that provides oxygen to the rest of the root system

auxin: a plant hormone that regulates growth

capillary water: water that is held in small soil spaces

cork cambium: layer of cells found in the outer bark of a woody stem that produces cork tissue

cortex: storage tissue found in roots and stems

dormant: in a resting condition in which growth stops and metabolism is minimal

endodermis: single layer of cells found at the inner boundary of the cortex of a plant root

epidermis: outer layer of cells in plant roots, stems, leaves, flowers, and seeds

fibrous root: member of a system that contains many branching roots

gibberellin: growth-regulating substance that promotes cell elongation and division

geotropism: growth response to gravity

herbaceous plant: plant with mostly primary tissues; usually only lives for one growing season

humus: organic matter found in soil composed of dead or decaying organisms

loam: type of rich soil

macronutrient: mineral that a plant needs in large amounts

micronutrient: mineral that a plant needs in small amounts

osmosis: the diffusion of water through a selectively permeable membrane from an area of high concentration to an area of lower concentration

pericycle: ring of parenchyma cells found around the vascular cylinder of a primary root

phloem: tissue in roots, stems, and leaves that conducts dissolved food particles

primary growth: growth in the length of a root caused by the formation of primary tissue

primary root: the first root that is pushed down into the soil from the lower end of the plant embryo

primary tissue: the first tissue that develops in a young root that adds length to the root

prop root: root that is pushed into the ground and helps the underground root system to support the plant

region of cell division: area of cells in the root apical meristem

region of elongation: area in the root in which cells become two to three times longer and slightly wider than their original size

region of maturation: area in the root in which cells mature and differentiate into specialized tissues

root: single structure that is part of a system that anchors a plant to the ground, absorbs water and nutrients for plant growth, and stores food and water

root cap: tissue located at the root tip that serves to protect the tissue behind it

root hair: tiny, finger-like projection found in young roots

secondary growth: an increase in the diameter of the root caused by the addition of secondary tissue

secondary tissue: root tissue produced by the vascular cambium that develops after the primary root has formed

taproot: the major root of a plant

transpiration: the evaporation of water from the leaves of a plant

tropism: movement stimulated by external factors such as light or gravity

vascular cambium: the meristem that forms between xylem and phloem in the vascular cylinder

woody plant: plant that has secondary xylem as the major part of its stem and roots

xylem: the woody tissue of a root or stem that conducts water and dissolved minerals upward in the plant

Name _____ Date _____

Class _____

ROOTS:
BASIC TERMS CROSSWORD PUZZLE

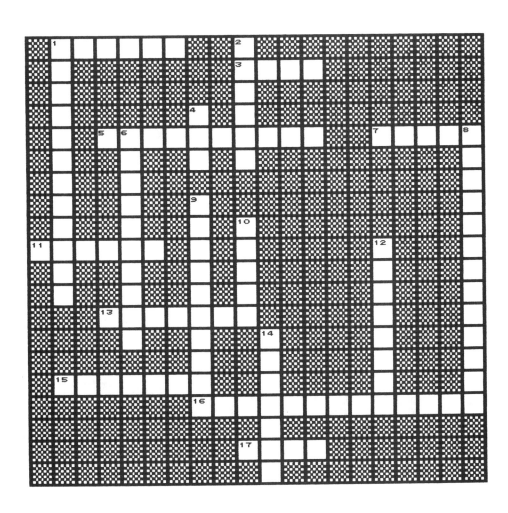

ACROSS

1. _____ transport requires the use of energy
3. A root _____ is a thin projection from young root cells
5. Plant with mostly primary tissues; usually lives only one year
7. Tissue that conducts materials from roots to other parts of plant
11. Storage tissue in roots and stems
13. _____ roots have many branches
15. Movement of water through a semipermeable membrane, from high concentration to low
16. Plant needs a lot of this
17. _____ cambium produces cork tissue

DOWN

1. _____ roots develop from stem or leaf nodes
2. Conducts dissolved foods
4. Root _____ protects root tissue
6. Layer of cells at inner boundary of cortex
8. Plant needs a small amount of this
9. Growth response to gravity
10. Organic matter in soils
12. The _____ cambium is meristem between xylem and phloem
14. Major root

ROOTS:
ADVANCED TERMS CROSSWORD PUZZLE

ACROSS

2. Evaporation of water from leaves
6. The _____ root provides oxygen to the rest of the roots
7. Ring of parenchyma cells around the vascular cylinder of root
8. The _____ root is a large main root
9. Conducts materials downward
13. Region of _____ is where root cells become longer
16. _____ plants generally live more than one year
17. Plant needs only a little of this
18. _____ roots develop from the node of a stem or leaf

DOWN

1. Region of _____ is where the root cells differentiate into specialized tissues
3. _____ roots help support the plant
4. The root _____ protects the other cells behind it
5. Tissue conducts material upward
10. Growth response to gravity
11. _____ water is in tiny spaces in the soil
12. _____ plants generally live only one year and do not develop woody stems
14. Movement of water through a semipermeable membrane from a region of high concentration to low
15. Vascular _____ cells produce xylem and phloem cells

Name _____   Date _____

Class _____

ROOTS:
VOCABULARY WORDSEARCH

The wordsearch below contains terms related to our study of roots. The words can be found horizontally in either direction, vertically in either direction, and diagonally in either direction. Clues are given to help you find the words.

```
Y P C A P I L L A R Y F G J N O A Y
I M H M A G S F R C T T N A M R O D
K B F E X R U N L O N E P P C F A W
E X C R I R O A K R Q A O N R U I O
L L A O F M R U A T O Q I O X A A S
O D M X U L B A M E V E S I S U D M
N A B C T S I O I X D S N T D Z T O
G V I P G U F S C P H K T A D U S S
A N U E I O H L R R V S V R Q J W I
T Z M R B E L U O N F O H I J U R S
I U H I B C D K N B T H K P D Z L K
O E H C E A P R U X V U L S U F A H
N C F Y R B Y J T B N M I N U W I C
L W H C E R Z W R X T U Q A C E R R
B O L L E R V I L I S A R N C E S
K X M E L H I S E R U C Q T Y K A H
H U E J I I C A N H M E O L H P Z F
P B L N N U X H T P W R H O H A N X
Y Q Y S F Z S T R O P I S M C U Y D
W K X L S Z M H A B H R S S X R H J
```

CLUES

1. A _____ root has many branches
2. Ring of parenchyma cells around vascular cylinder of primary root
3. Cork _____ produces cork tissue
4. _____ water is held in small soil spaces
5. Mineral needed in small amounts
6. Growth-regulating chemical that promotes growth
7. Root cells become much longer in the region of _____
8. Tissue conducting materials upward
9. _____ plants usually live only one growing season and are not woody
10. Loss of water from the leaves of a plant
11. _____ roots provide oxygen to the rest of the root system
12. Movements stimulated by external factors such as light
13. Storage tissue in roots and stems
14. Tissue that conducts dissolved food downward
15. Organic matter in soil
16. Diffusion of water through a semipermeable membrane from an area of higher concentration to a region of lower concentration
17. Plant's growth-regulating hormone
18. Resting or "sleeping"

VOCABULARY

Basic		*Advanced*
annual ring	pith	(all of the basic vocabulary)
bark	rhizome	adhesion
bud scale	sapwood	auxin
bulb	stolon	cohesion
corm	terminal bud	epicotyl
cortex	transpiration	gibberellin
epidermis	vascular bundle	ground tissue
heartwood	vascular cambium	lenticel
lateral bud	wood	phototropism
leaf scar	woody stem	plumule
phloem	xylem	turgor

DEFINITIONS

adhesion: the attraction of water molecules to the walls of vessels that helps support the water column in plants

annual ring: the growth of secondary xylem consisting of springwood and summerwood during one year

auxin: plant growth-regulating hormone

bark: all tissues, including cork, cork cambium, cortex, and phloem, on the outside of trees

bud scale: modified leaf that protects a bud

bulb: underground stem composed of a central stem and a bud surrounded by thick leaves

cohesion: the attraction between like molecules

corm: shortened underground stem that has leaves reduced to thin scales

cortex: storage tissue in roots and stems

epicotyl: the part of the plant embryo that is located above the cotyledons

epidermis: outer layer of cells in roots, stems, leaves, flowers, and seeds

gibberellin: growth-regulating substance that promotes cell elongation and division

ground tissue: in conifers and dicots, tissue consisting of the inner pith and outer cortex; in monocots, it is not separated into pith and cortex

heartwood: the dark region of an older stem that no longer conducts water

lateral bud: bud that forms in a leaf axil

leaf scar: mark that shows the location of a node or place of attachment of a leaf

lenticel: opening in the cork layer through which gases are exchanged

phloem: tissue in stems that conducts dissolved food substances

phototropism: the growth of plants toward light

pith: inner part of stems in gymnosperms and dicots

plumule: part of a plant embryo that consists of the first pair of leaves

rhizome: underground stem that is horizontal, stores food, and can propagate vegetatively

sapwood: the light part of a stem through which water still travels

stolon: slender stem that grows horizontally above the ground

terminal bud: bud at the tip of a stem

transpiration: the process in which water evaporates from the leaves of plants

turgor: pressure that builds up in a cell as water travels in through osmosis

vascular bundle: xylem and phloem cells in the stems of conifers and dicots

vascular cambium: the meristem that forms between the xylem and phloem in the vascular cylinder

wood: the xylem cells in a plant

woody stem: stem containing bark, vascular cambium, wood, and pith

xylem: stem tissue that conducts water and dissolved minerals upward

Name _____   Date _____

Class _____

STEMS:
BASIC TERMS CROSSWORD PUZZLE

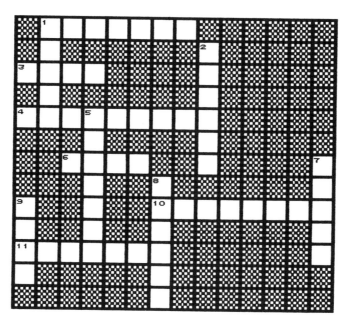

ACROSS

1. Part of woody stem through which water still travels; light in color
3. Outer tissues of tree
4. Dark region of older stem; no longer conducts water
6. Inner part of gymnosperm or dicot stem
10. The _____ bud is at the tip of a stem
11. The _____ bud forms in a leaf axil

DOWN

1. The bud _____ protects a bud
2. Xylem and phloem are found in the vascular _____
5. Horizontal underground stem
7. Conducts water and minerals upward
8. Horizontal aboveground stem
9. Underground stem with a bud surrounded by thick leaves

Name _____ Date _____

Class _____

STEMS:
ADVANCED TERMS CROSSWORD PUZZLE

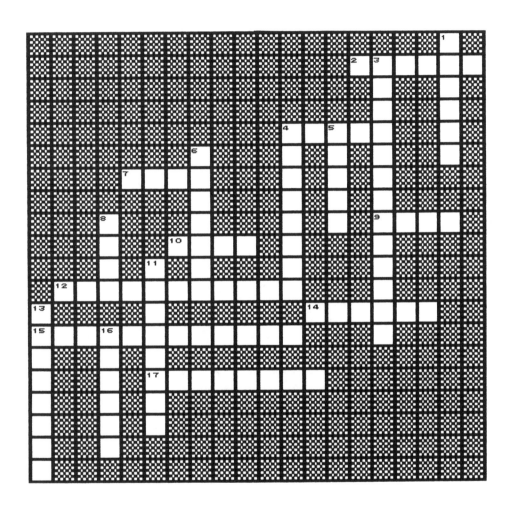

ACROSS

2. Horizontal aboveground stem
4. Plant growth-regulating hormone
7. Inner part of gymnosperm or dicot stem
9. Annual _____ shows one year's growth
10. Short underground stem
12. Growth-regulating substance
14. Conducts downward
15. Plant growth response to light
17. Attraction between like molecules

DOWN

1. Storage tissue
3. Evaporation of water from leaves
4. Attraction of water molecules to walls of vessels
5. Conducts upward
6. Horizontal underground stem
8. Underground stem; bud surrounded by thick, leaves
11. Opening in cork layer for gas exchange
13. Part of plant embryo above the cotyledons
16. Pressure due to fluids

STEMS:
VOCABULARY WORDSEARCH

The wordsearch below contains terms related to our study of stems. The words can be found horizontally in either direction, vertically in either direction, and diagonally in either direction. Clues are given to help you find the words.

```
M C O S I M R E D I P E Y P R G P J Z N
J I Q X Z R I E H M P L V F J H E J Q U
S T N E O A E B C W X P B Z O Y T V N E
U S M G V K G D H H F G Q T B I A T O M
J N R T X Y T F R T W O O I G I A B I Q
I U O V S N B C L Y U T R V T B S I T X
T O X I C K E R C E R E N J S E K X A S
O X Z I S M Z F D O Q R Z I P N C L R W
L N T X N E N D P E Z D X J B O C R I I
Y A F L A F H I F Q M W M A J A W M P F
T S H C V Y S D O G F C Z H M U E R S F
O P E D M M U W A X Q Y G B X G P T N H
C A R R Y C V K Q X L Q I A Y M C C A J
I U A H X L O W G Z D U J J P C N E R E
P X R J I U N R G C M Z V O H M I C T Y
E I N X U Z V F T M Z Y N U L L L A Q T
N N Y G R V O E R E L Z G U O A U X M W
E M E M O L T M E L X C K X E S Z S F Z
M M R D O A Q A E Y H J X B M G N S L H
J R F H T U N I K X Z A R M R W W H U D
```

CLUES

1. Water pressure in a cell
2. Attraction of water molecules to walls of vascular tissue
3. Tissue that conducts dissolved foods
4. Horizontal underground stem
5. Growth response to light
6. Plant growth-regulating hormone
7. The vascular _____ produces the xylem and phloem cells
8. Outer cell layer
9. Part of plant embryo above the cotyledons
10. Conducts water and dissolved minerals upward
11. Evaporation of water from leaves
12. Storage tissue in stems and roots

15: LEAVES

VOCABULARY

Basic

abcission
blade
bundle sheath
compound leaf
deciduous
epidermis
guard cell
leaf
leaflet
mesophyll
needle

palisade layer
petiole
simple leaf
spine
spongy layer
stoma
stomata
tendril
touch response
turgor pressure
venation

Advanced

(all of the basic vocabulary)
bipinnately compound
palmate venation
palmately compound
parallel venation
parenchyma
photoperiodism
pinnate venetion
pinnately compound
sessile leaf

DEFINITIONS

abcission: the separation of a leaf from a plant

bipinnately compound: leaves that are twice-divided

blade: broad flat part of a leaf

bundle sheath: layer of cells surrounding a leaf vein

compound leaf: leaf divided into separate parts called leaflets

deciduous: woody plants that lose their leaves at the end of the growing season

epidermis: outer layer of cells in leaves

guard cell: bean-shaped cell located around the stomata

leaf: the plant organ in which photosynthesis occurs

leaflet: one of the divisions found in a compound leaf

mesophyll: tissue between the upper and lower epidermis of a leaf

needle: long, stiff leaf found, for example, in pine trees

palisade layer: tissue of leaf mesophyll that consists of elongated cells containing chloroplasts

palmate venation: leaf venation in which the veins spread out in a net-like pattern

palmately compound: leaf with leaflets attached to a single petiole

parallel venation: the arrangement of leaf veins found in monocot plants in which there are several parallel veins of approximately the same size

parenchyma: tissue made up of thin-walled, loosely packed cells that form cortex and pith

71

petiole: the stalk by which a leaf is attached to a stem

photoperiodism: response in plants to changing periods of light and darkness

pinnate venation: leaf venation in which one large vein runs the length of the blade with small veins branching from it to the edge

pinnately compound: leaves that have leaflets attached along both sides of the petiole

sessile leaf: leaf without a petiole

simple leaf: leaf with one blade and one petiole

spine: specialized leaf of a cactus adapted to a desert environment

spongy layer: tissue of the leaf mesophyll found between the palisade layer and the lower epidermis

stoma: leaf pore that regulates the passage of gases and water vapor

stomata: plural of stoma

tendril: specialized leaf that wraps around a solid object and helps to support a climbing plant

touch response: quick movement of plant parts when they are touched

turgor pressure: pressure that builds up in cells as water travels through osmosis

venation: the arrangement of veins in a leaf

Name _____ Date _____

Class _____

LEAVES:
BASIC TERMS CROSSWORD PUZZLE

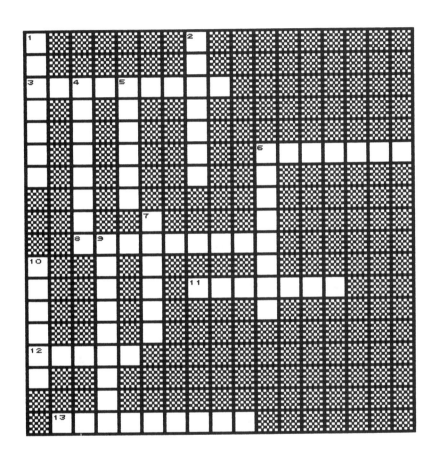

ACROSS

3. Separation of leaf from plant
6. Leaf stalk
8. Plant that loses leaves
11. Leaf specialized for climbing
12. _____ cells surround stomata
13. Tissue between upper and lower epidermis of leaf

DOWN

1. Division of compound leaf
2. Openings in leaf; allow passage of water vapor and gases
4. The _____ leaf is divided into leaflets
5. The bundle _____ surrounds the leaf vein
6. The _____ layer consists of elongated mesophyll cells and is the site of most photosynthesis
7. _____ leaves have one blade and one petiole
9. Outer layer of cells
10. The _____ layer is between the lower epidermis and the palisade layer

LEAVES:
ADVANCED TERMS CROSSWORD PUZZLE

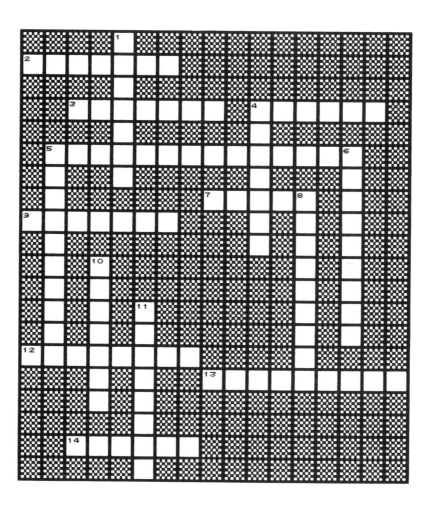

ACROSS

2. A _____ leaf has no petiole
3. Leaf modified for climbing
4. _____ layer is between the lower epidermis and the palisade layer
5. Plant response to changing periods of light and darkness
7. _____ cells surround the stomata
9. Leaf stalk
12. _____ venation is usually found in monocot leaves
13. Layer of separation of leaf from plant
14. _____ sheath surrounds the vein

DOWN

1. _____ venation has one large vein running the length of the leaf with small veins branching off
4. Openings for gases and water vapor
5. Cortex and pith; thin-walled cells
6. Tissue between upper and lower epidermis of leaf
8. Woody plants that lose leaves
10. _____ leaves have veins spreading outward
11. The _____ layer of the mesophyll has elongated cells with chloroplasts

Name _____ Date _____

Class _____

LEAVES:
VOCABULARY WORDSEARCH

The wordsearch below contains terms related to our study of leaves. The words can be found horizontally in either direction, vertically in either direction, and diagonally in either direction. Clues are given to help you find the words.

```
A H A T X Q K C L Z D Z Q T U R G O R S
P L P G U A R D C E L L E T P L R J W U
N Y P I X T X V M M C M W T S Z Q N A H
Z Z H N O I T A N E V A R V A H C S X A
H R O L S O E E H F S U P B H N R D Q T
W P T P A B C I S S I O N T R Y N R D A
X A O B A U D I H A V D P K B Y W I L M
B L P W A R X V Q B B K K H I E X P P O
W I E P I D E R M I S A A H Y M L C J T
G S R R L Q I N X G A V R E H L P O N S
X A I Y F Z N X C P J L D L O Q L Y G I
I D O R Y Z Q C E H F O B I V V S W V M
B E D Q I S C T P Q Y O O S W A E W P L
N O I K L H I J B O S M Y S N E E O P I
F O S S U O U D I C E D A E W T I J N R
M C M B L K O Q U C D F R S A R A S I D
K Y R E W X O F R K O H K M U H H O I N
Y L W E O F F F U D Z N L C N M H B A E
V R T Y Y O K E R Y U A A V O Q J H I T
X Z O N B Z F O U P P M M L J W U J J H
```

CLUES

1. A _____ leaf has no petiole
2. Response in plants to changing periods of light and darkness
3. _____ pressure increases in cells as water enters the cells
4. Leaf stalk
5. Cells found surrounding stomata (two words)
6. Separation of leaf from plant
7. _____ venation describes a leaf with one main vein running the length of the blade with small veins branching from it to the edge
8. A _____ leaf has several veins radiating from one point

9. Specialized leaf; helps support a climing plant
10. The _____ layer consists of elongated cells containing chloroplasts
11. Outer layer of cells
12. Thin-walled loosely packed cells in cortex and pith
13. Arrangement of veins
14. Leaf pores
15. Tissue between upper and lower epidermis of a leaf
16. Woody plants that lose leaves

VOCABULARY

Basic		*Advanced*	
annual	ovary	(all of the basic	multiple fruit
anther	ovule	vocabulary)	micropyle
biennial	perennial	aggregate fruit	microspore
calyx	petal	berry	nut
carpel	pistil	capsule	plumule
corolla	pollen grain	drupe	pod
cotyledon	pollination	endosperm nucleus	polar nucleus
cross-pollination	receptacle	epicotyl	pome
dormancy	seed	generative cell	radicle
dry fruit	seed coat	grain	tube cell
fleshy fruit	sepal	hilum	winged fruit
flower	stamen	hypocotyl	
fruit	stigma	megaspore	
germination	style		
nectar			

DEFINITIONS

aggregate fruit: mass of small simple fruits, each formed from one ovary

annual: plant that completes its life cycle in one growing season

anther: the part of the stamen that produces pollen grains

berry: succulent fruit formed from a single flower with several fused carpels and containing one or several seeds

biennial: plant that takes two growing seasons to complete its life cycle

calyx: the outer ring of flower parts containing all the sepals of a flower

capsule: dry fruit formed from one or more fused carpels and containing many seeds

carpel: a leaf-like floral structure enclosing the ovule or ovules of angiosperms

corolla: all the petals of a flower

cotyledon: the seed leaf of a plant

cross-pollination: the transfer of pollen from the anthers of one plant to the stigmas of another plant of the same species

dormancy: resting condition of plants in which growth stops and metabolism is minimal

drupe: succulent fruit formed from one carpel and containing one seed

dry fruit: fruits including capsules and pods that open when ripe

endosperm nucleus: triploid nucleus formed by the fusion of a sperm cell with two polar nuclei

epicotyl: the part of a plant embryo located above the cotyledons

fleshy fruit: type of fruit that includes pomes, drupes, and berries

flower: the part of some plants that bears the reproductive structures

fruit: ripened ovary that encloses a seed or seeds

generative cell: one of the two haploid cells found in a pollen grain

germination: the beginning growth of a seed, spore, bud, or other reproductive structure

grain: single small, hard seed

hilum: the oval scar on a seed where it is attached to the pod wall

hypocotyl: the part of the stem of a seedling that lies below the point of attachment of the stalk of the cotyledon and above the primary root

megaspore: large spore that develops into a female gametophyte

multiple fruit: type of fruit formed from an inflorescence, for example, pineapple, fig, or mulberry

micropyle: opening in the ovule wall through which the pollen tube enters

microspore: small spore that develops into a male gametophyte

nectar: solution of sugar and water produced by many plants

nut: dry, one-seeded fruit in which the pericarp forms a hardy, woody shell

ovary: organ that produces female gametophytes

ovule: the structure in the ovary of a flower that contains the female gametophyte

perennial: plant that takes many years to reach full size and begin to reproduce

petal: part of the second ring of floral parts, often brightly colored

pistil: a single carpel or a group of fused carpels

plumule: the part of the plant embryo that consists of the first pair of leaves

pod: dry fruit formed from a single carpel that includes peas and beans

polar nucleus: one of two nuclei in the embryo sac that fuses with one of the sperm nuclei to form the endosperm nucleus

pollen grain: structure produced by the anther of a flower that contains the male sex cells

pollination: the transfer of pollen from an anther to a stigma

pome: type of fruit that develops from the receptacle of the flower, and which becomes swollen and fleshy

radicle: the embryonic root in a seed

receptacle: the end of a flower stalk that bears the reproductive structures

seed: plant embryo surrounded by an endosperm and protected by a seed coat

seed coat: the enternal protective covering of a seed that is usually hard and dry

sepal: the outermost ring of flower parts that protects and encloses the bud

stamen: the male reproductive part of the flower

stigma: the tip of the pistil

style: slender, stalk-like structure on which the ovary rests

tube cell: one of the two cells produced by a dividing microspore in a pollen grain

winged fruit: dry fruit with prominent wings attached to the ovary wall

FLOWERS:
BASIC TERMS CROSSWORD PUZZLE

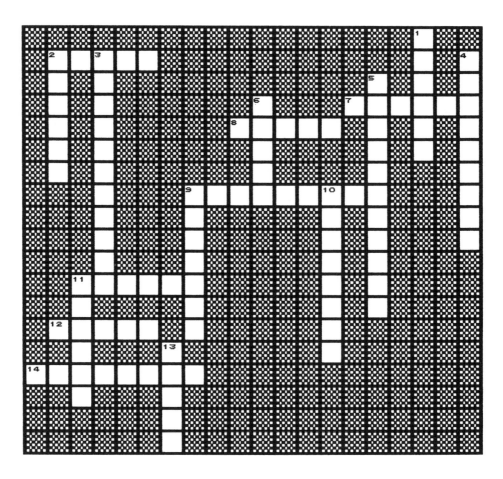

ACROSS

2. Protects and encloses flower bud
7. Sugary solution
8. In ovary, contains female gametophyte
9. Seed leaf
11. Completes life cycle in one year
12. Stalk on which ovary is found
14. Completes life cycle in two years

DOWN

1. Female reproductive part
2. Male reproductive part
3. Transfer of pollen from anther to stigma
4. Plant that takes several years to complete life cycle
5. Beginning growth of seed
6. Produces female gametophytes
9. All petals as a unit
10. Period of minimal metabolism
11. Produces pollen
13. All sepals

Name _____ Date _____

Class _____

FLOWERS:
ADVANCED TERMS CROSSWORD PUZZLE

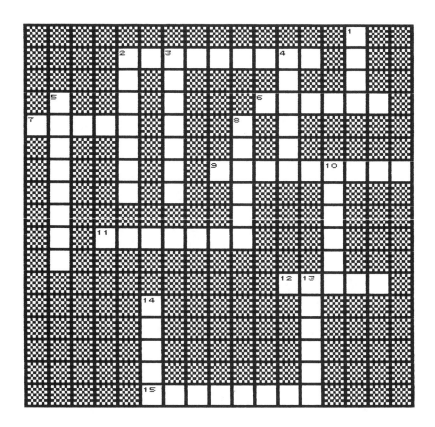

ACROSS

2. Takes many years to complete life cycle
6. Male part of a flower
7. Scar where seed was attached to pod wall
9. Opening in ovule wall
11. All petals
12. All sepals
15. Part of embryo located above cotyledons

DOWN

1. Swollen, fleshy fruit that develops from receptacle
2. First pair of leaves
3. Embryonic root
4. Produces pollen
5. Completes life cycle in two years
8. Tip of the pistil
10. Female flower part
13. Completes life cycle in one year
14. Supports ovary

FLOWERS:
VOCABULARY WORDSEARCH

The wordsearch below contains terms related to our study of flowers. The words can be found horizontally in either direction, vertically in either direction, and diagonally in either direction. Clues are given to help you find the words.

```
I H V I M T M R R B F Z I J H H C K Q R
I N U F I O B K L M I E O E C D R D Q G
Y J L C U V K Z K A B N E L K H Y M X N
O G E L A U N N A O P B M R F P E G Z H
B T K S O I X G I N A E L U R V T V R T
Z C H S J E J I Q C I Y S F Z W J J S L
T D P I S T I L Y P T C X Q K P Z X M O
Q H Z H H L A E L O G U N L E L Y T S B
I V D C E X M J C Y W E L Y P O R C I M
O Z R S I O G O X H T W O V Y I C N E Y
O P U G P F P N H Y L O R Y U U O J E V
P M P T T Y Y L R Q Q M C T U I U C F H
A Q E E H P P M U L I H L I T L W P V Y
L L A I N N E R E P W D P A P Z R A Q C
L Y L L E C E B U T E V N E E E M L R E
O Q J M Z C P F W L M I V S T G T E O D
R Q M E B I T F U X M E G M I A H X Z E
O N E M A T S V O R E J H T I T L O D I
C K X U P C O Y E S J K S K N X R Z L K
W H A L X K V G S T R J P A V R F K L G
```

© 1991 by Center for Applied Research in Education

CLUES

1. Plant that completes life cycle in one growing season
2. Beginning growth of a seed
3. Part of second ring of floral parts, often brightly colored
4. Male reproductive part of flower
5. Succulent fruit formed from one carpel and containing one seed
6. Part of the stem of a seedling, below the stalk of the cotyledon and above the primary root
7. Cell produced by a dividing microspore (two words)
8. Produces pollen
9. Structure in ovary that contains female gametophyte
10. Female reproductive part of flower
11. Tip of pistil
12. Plant embryo above cotyledons
13. Opening in ovule wall
14. All petals
15. Plant that takes several years to reach maturity
16. Protective outermost floral ring
17. Ovary stalk
18. Scar on a seed indicating where it was attached to the pod wall
19. Fruit that develops from the receptacle

VOCABULARY

Basic		Advanced
amebocyte	motile	(all of the basic vocabulary)
anterior	multicellular organism	basal disk
asymmetrical	nematocyst	blastopore
bilateral symmetry	nerve net	blastula
budding	osculum	calcium carbonate
Coelenterata	phylogenetic tree	central axis
coelom	polyp	coelenteron
collar cell	Porifera	gastrodermis
coral	posterior	gastrula
dorsal	radial symmetry	gemmule
ectoderm	regeneration	germ layer
endoderm	sessile	metazoan
excurrent pore	specialization	polarity
flagellum	spherical symmetry	spongin
gastrovascular cavity	spicule	
hydra	sponge	
incurrent pore	stinging cell	
interdependence	symmetry	
invertebrate	tentacle	
medusa	ventral	

DEFINITIONS

amebocyte: amoeba-like cell in sponges that absorbs food and excretes wastes

anterior: describes the head or front of an animal

asymmetrical: animal body plan with irregular form

basal disk: structure in sponges that allows attachment to surfaces

bilateral symmetry: body plan in animals in which the animal can be halved in one plane to form mirror images

blastopore: indentation that forms in a blastula

blastula: early stage of embryonic development in which cells divide to form a hollow sphere

budding: type of asexual reproduction found among simple organisms

calcium carbonate: substance that forms the spicules of some sponges; lime

central axis: in radially symmetrical animals, the line that passes through the center of the animal

Coelenterata: phylum of the animal kingdom containing hydras, corals, and sea anemones

coelenteron: the hollow body cavity of a coelenterate

coelom: body cavity surrounded by mesoderm tissue; it contains many organs

collar cell: flagellated cell found in the inner layer of sponges

coral: colonial coelenterate that forms a hard skeleton made of lime

dorsal: describes that part of an animal's body situated near or on the back, or lying towards the upper surface

ectoderm: layer of cells from which the skin and nervous system develops

endoderm: layer of cells from which the linings of the digestive system, liver, lungs, and other internal organs develop

excurrent pore: hole in the body of a sponge through which water is forced out

flagellum: thin, whip-like structure on the outside of cells that provides motility

gastrodermis: the innermost layer of cells in the bodies of coelenterates

gastrovascular cavity: the hollow body cavity found in coelenterates

gastrula: the stage in embryo development during which the primary germ layers form

gemmule: bud formed by coelenterates that can survive unfavorable conditions

germ layer: layer of cells in a developing organism that gives rise to specific structures

hydra: small, sessile coelenterate with tentacles for obtaining food

incurrent pore: small hole in a sponge into which water passes

interdependence: the dependence of specialized cells of multicellular organisms on other cells

invertebrate: animal that does not have a backbone

medusa: bell-shaped, free-swimming form in the coelenterates

metazoan: multicellular organism

motile: having the ability to move

multicellular organism: living thing that consists of more than one cell

nematocyst: stinging cell in coelenterates

nerve net: primitive type of nervous system that allows coelenterates to react to stimuli

osculum: large pore in the central cavity of sponges through which water leaves

phylogenetic tree: diagram that represents a hypothesis as to the origin and evolution of various groups of organisms

polarity: of an organism, having two different ends

polyp: a form with a tubular body that has a basal disk at one end and tentacles at the other end

Porifera: phylum in the animal kingdom that includes sponges

posterior: part of the animal furthest away from the head or front

radial symmetry: body plan in which an animal can be divided through many planes along its axis to form mirror images

regeneration: ability of certain animals to regrow body parts

sessile: of animals, living attached to another object

specialization: the adaptation of a cell for a particular function

spherical symmetry: an organism that is shaped like a ball and has cells that are all alike

spicule: silicon or lime structure that forms the skeleton of certain sponges

sponge: animal that belongs to the phylum Porifera

spongin: tough, flexible protein that makes up the fibrous network which supports some sponges

stinging cell: structure found in the tentacles of some coelenterates used to capture prey

symmetry: correspondence in shape on opposite sides of a plane, or axis

tentacle: long appendage of certain coelenterates that have stinging cells

ventral: the part of an animal lying on or near its lower surface

SPONGES AND COELENTERATES:
BASIC TERMS CROSSWORD PUZZLE

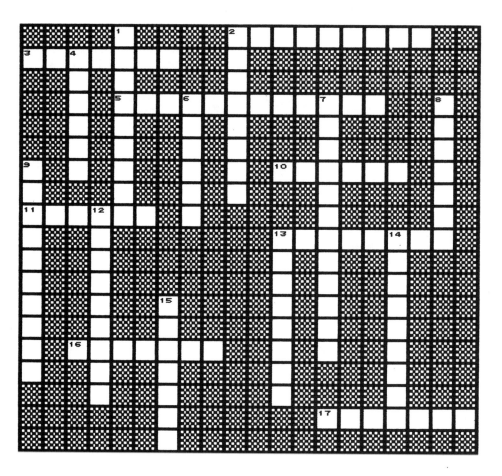

ACROSS

2. Amoeba-like cell in sponges
3. Asexual reproduction
5. Having irregular, unbalanced form
10. _____ cells are flagellated and found on a cell's inner layer
11. Able to move
13. _____ cells develop into skin and nervous system
16. Unable to move
17. On or towards lower surface

DOWN

1. Can be halved to form mirror images
2. Towards the front
4. Towards the upper surface
6. Bell-shaped, free-swimming form in coelenterates
7. Phylum containing hydras, corals, and sea anemones
8. Large opening in sponges
9. Stinging cell
12. _____ pores allow water to enter sponges
13. Layer of cells that develops into digestive system, lungs, and liver
14. _____ pores allow water to leave sponges
15. Sponge skeleton structure

Name _____ Date _____

Class _____

SPONGES AND COELENTERATES:
ADVANCED TERMS CROSSWORD PUZZLE

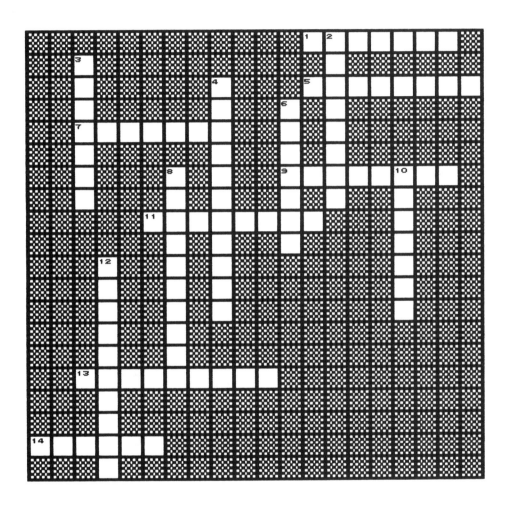

ACROSS

1. Flexible sponge skeleton
5. Embryonic stage, hollow sphere
7. Hard sponge skeleton
9. Multicellular organism
11. Embryonic stage; primary germ layers form
13. Away from head
14. Upper surface

DOWN

2. Condition of having two different ends or sides
3. Not moving
4. Hollow body cavity of coelenterate
6. Coelenterate bud
8. Indentation in blastula
10. Large opening to center of sponge
12. Stinging cell

SPONGES AND COELENTERATES:
VOCABULARY WORDSEARCH

The wordsearch below contains terms related to our study of sponges and coelenterates. The words can be found horizontally in either direction, vertically in either direction, and diagonally in either direction. Clues are given to help you find the words.

```
D A S R O F K X U J I C T R T X L K G A
Y N C E A I K X O D G W M X Z O Q V C Q
A T G F L A G E L L U M E L U M M E G J
X E Q I G Z V A P O R I F E R A U A D T
E R S A M E B O C Y T E T H M M R Q O A
L I E R O P O T S A L B R E Z N O J R J
I O Z N C C R C P W F B B Q N E I N S S
S R L S O O M G F N H K W T A T R J A F
S M G P S L I C B T D U T F O P E L L F
E Q A O G L U L E J N Z S E Z J T Q B S
S B S N B A L E R Z F N Y W A T S L C S
V X T G V R N V K R X N C C T K O A O B
A R R I G C W D K Z S W O H E U P Y O Q
L L U N N E L E W L P M T E M D N O E N
C T L G I L R M N K I C A L L I G L H H
W M A I D L C J W L C J M U H I N B F S
M H T V D Q T R G L U M E W G N T R U I
Y C Y B U X B V O O L H N I V W W O Y E
W E E N B I M A T Q E F L F O N A F M H
D G V Q I M E V O I H R D F N P I L F F
```

CLUES

1. Sponge supporting material made of flexible protein
2. Embryo developmental stage in which primary germ layers form
3. Nonmoving; attached
4. Phylum of sponges
5. Towards upper surface or back
6. Towards head or front
7. Multicellular animal
8. Indentation in blastula
9. Able to move
10. Stinging cell
11. Flagellated cell inside a sponge (two words)
12. Amoeba-like cell in sponges
13. Bud formed by coelenterates; helps species survive
14. Silicon or lime sponge skeleton
15. Away from front or head
16. Whip-like structure that aids locomotion
17. Type of asexual reproduction

VOCABULARY

Basic

Annelida	mesoderm
aortic arch	nematode
aquatic	open circulatory
closed circulatory	system
system	parasite
coelom	pharynx
crop	photosensitive
cyst	Platyhelminthes
esophagus	proglottid
eyespot	rotifer
flame cell	roundworm
fluke	scolex
ganglion	segment
gizzard	segmented worm
hermaphrodite	tapeworm
hookworm	terrestrial
larva	trichina worm
marine	uterus

Advanced

(all of the basic vocabulary)
acoelomate
ascaris
cephalization
Cestoda
clitellum
cuticle
eucoelomate
genital pore
longitudinal nerve
nephridium
planarian
prostomium
pseudocoelom
seminal receptacle
seminal vesicle
setae
transverse nerve
Trematoda
Turbellaria

DEFINITIONS

acoelomate: animal that lacks a coelom

Annelida: phylum of the animal kingdom that includes the earthworm and other segmented worms

aortic arch: structure in the earthworm that keeps blood flowing through its body

aquatic: living in water

ascaris: large roundworm that can live in the intestines of pigs, horses, and humans

cephalization: the concentration of nerves and receptors at the anterior end of an organism

Cestoda: class of parasitic flatworms that includes the tapeworm

clitellum: structure on the earthworm that swells during the reproductive cycle

closed circulatory system: circulatory system in which the blood stays within specific vessels

coelom: body cavity surrounded by mesoderm and containing many organs

crop: food-storing organ in the alimentary canal in earthworms

cuticle: thick, non-living layer secreted by the epidermis of some parasites that serves as a means of protection

cyst: hard-walled body formed by some invertebrates that protects the organism from unfavorable conditions

esophagus: the food tube that connects the mouth and the stomach

eucoelomate: animal with a true coelom

eyespot: simple photosensitive organ

flame cell: excretory organ that collects excess water and cell wastes

fluke: common parasitic flatworm found in many animals

ganglion: mass of neurons located outside the central nervous system

genital pore: opening through which the eggs of the fluke are released

gizzard: organ that grinds food in the digestive system of some animals

hermaphrodite: organism that has the reproductive organs of both sexes

hookworm: parasitic roundworm that enters the body by boring holes in the feet of its host

larva: the immature stage in the life cycle of some animals

longitudinal nerve: nerve that runs the length of the body in planarians

marine: living in the ocean

mesoderm: the middle layer of cells in most animal embryos

nematode: parasitic roundworm

nephridium: excretory structure found in worms

open circulatory system: type of circulatory system in which blood bathes the body organs directly

parasite: an animal that lives on or in, and obtains foods and sometimes protection from, another living organism

pharynx: the food tube in planarians

photosensitive: describes an organ that responds to light

planarian: a free-living flatworm of the class Turbellaria

Platyhelminthes: phylum in the animal kingdom that includes the flatworms

proglottid: segment of the tapeworm's body

prostomium: the upper lip in an earthworm

pseudocoelom: the body space between the endoderm and the mesoderm

rotifer: small, parasitic marine organism with a crown of cilia at its anterior end

roundworm: any member of the phylum Nematoda, including the ascaris, hookworm, and trichina worm

scolex: knob-shaped head with hooks or suckers

segment: section of a worm's body

segmented worm: worm with a body separated into many sections

seminal receptacle: structure that receives sperm cells

seminal vesicle: structure that stores sperm cells

setae: bristles used for locomotion

tapeworm: parasitic flatworm with a flat ribbon-like body

terrestrial: living on land

transverse nerve: crosswise nerve that connects the longitudinal nerves in planaria

Trematoda: class in the animal kingdom containing the flukes

trichina worm: parasitic roundworm that spends the first part of its life as a cyst in the muscles of pigs, dogs, cats, or rats

Turbellaria: class of flatworms containing planarians

uterus: long, coiled tube in which many eggs are stored

Name _____ Date _____

Class _____

ROTIFERS, FLATWORMS, ROUNDWORMS, AND SEGMENTED WORMS: BASIC TERMS CROSSWORD PUZZLE

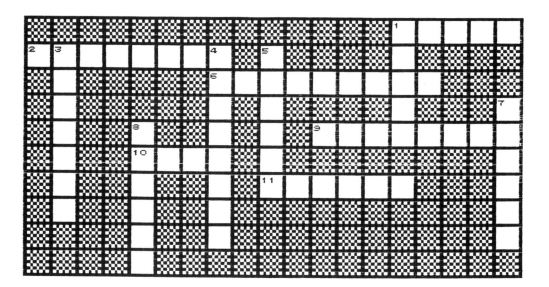

ACROSS

1. _____ cells are excretory organs
2. Cluster of nerve cells
6. Connects mouth to stomach
9. Phylum of segmented worms
10. Hard protective coating
11. _____ circulatory systems have the blood in blood vessels

DOWN

1. Parasitic flatworm
3. Living in water
4. Roundworm
5. _____ arch keeps the blood flowing through the body
7. Living in the ocean
8. Tapeworm head

Name _____   Date _____

Class _____

ROTIFERS, FLATWORMS, ROUNDWORMS, AND SEGMENTED WORMS: ADVANCED TERMS CROSSWORD PUZZLE

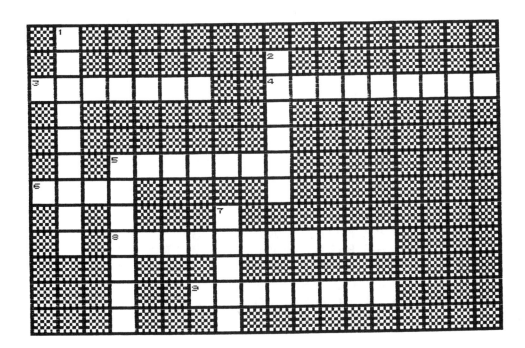

ACROSS

3. The seminal _____ stores sperm
4. Earthworm structure that swells during reproductive cycle
5. Thick secreted covering
6. The fluke's eggs are released through the genital _____
8. Class of flatworms that includes planarians
9. Body of neurons located outside the central nervous system

DOWN

1. Class containing flukes
2. "Head" of tapeworm
5. Class that includes the tapeworm
7. Bristles, usually used for moving

ROTIFERS, FLATWORMS, ROUNDWORMS, AND SEGMENTED WORMS: VOCABULARY WORDSEARCH

The wordsearch below contains terms related to our study of rotifers, flatworms, roundworms, and segmented worms. The words can be found horizontally in either direction, vertically in either direction, and diagonally in either direction. Clues are given to help you find the words.

```
K E Q L A O M M U L L E T I L C O L N Z
M N K D R N D C U W P H U H V T V U M X
A Q C S A N N P X R R Q L E H D R G Z Z
C M L V I O L E W X N Y S L A P M P Q P
U B E W R H R B L Z W P N C Y M O U E R
T S Q G A N D P T I G M S I T V V P Q O
I Q H B L O Z H X K D J I S N S Z O K G
C W F T L I T V N N T A Z E V E Y B E L
L O P K E L A S P E H J X V B H E C E O
E E H R B G P U W V P D S L V J K X Q T
Z M M V R N E J E Q D H X A J W U T L T
I U J L U A W R P A S E R N E E L X W I
G O X A T G O S T S Q I T I A K F T W D
N O C Q G L R B T T D B B M D Z P B B W
V N U U K Y M A T V S P X E Z I E Z E Q
P A D A D O T A M E R T F S L I U Z N T
P B M T V V B P P N E M A T O D E M L R
G K K I S C O L E X C O E L O M V P H F
G C I C X K J W I H D V A F D O I Q E B
A C O F W A T Y B S F R G M H O D I D B
```

© 1991 by Center for Applied Research in Education

CLUES

1. Class of flukes
2. Stores sperm cells (two words)
3. Swells during reproductive cycle of earthworm
4. Tapeworm body segment
5. Type of parasitic flatworm
6. Phylum of segmented worms
7. Class of flatworms that contains planarians
8. Worm's excretory structure
9. Parasitic flatworm with flat, ribbon-like body
10. Roundworm

11. Resting stage with hard protective coating
12. Living in water
13. Bristles
14. Thick, non-living secreted protective layer
15. Tapeworm head
16. Group of neurons located outside the central nervous system

VOCABULARY

Basic

- bivalve
- echinoderm
- endoskeleton
- foot
- gastropod
- gill
- head
- mantle
- mantle cavity
- Mollusca
- nephridium
- nerve ring
- radula
- shell
- slug
- snail
- true coelom
- tube foot
- water-vascular system

Advanced

(all of the basic vocabulary)
- aboral surface
- bipinnaria larva
- cephalopod
- cleavage
- dermal branchia
- eucoelomate
- excurrent siphon
- horny layer
- incurrent siphon
- oral surface
- pearly layer
- pelecypod
- pentaradial symmetry
- prismatic layer
- trochophore
- veliger
- visceral hump

DEFINITIONS

aboral surface: the surface opposite the mouth in a starfish

bipinnaria larva: bilaterally symmetrical larva that develops from the fertilized egg of the starfish

bivalve: mollusk with two shells that are hinged together

cephalopod: large-headed mollusk with no shell or a small, reduced shell

cleavage: the division of one cell into two cells

dermal branchia: outpocketing of the coelom of the starfish that functions as both respiratory and excretory organ

echinoderm: radially symmetrical marine animal, such as the starfish, brittle star, sea urchin, and sea cucumber

endoskeleton: internal support system

eucoelomate: animal with a true coelom

excurrent siphon: in mollusks, the opening through which the water leaves the body

foot: large, muscular organ that mollusks use for locomotion

gastropod: "stomach-footed," single-shelled mollusk

gill: respiratory organ that absorbs dissolved oxygen from water

head: anterior part of the organism where sensory organs are usually located

horny layer: the thin outer layer of mollusk shells

incurrent siphon: the structure in a mollusk through which water passes into the body

mantle: in mollusks the outermost layer of the body wall or a soft extension of it, usually secretes a shell

mantle cavity: space formed by the mantle hanging down over the sides of the body of a mollusk

Mollusca: phylum of animals including snails, clams, oysters, and squid

nephridium: the excretory structure in mollusks

nerve ring: the basic nervous system structure of an echinoderm

oral surface: in echinoderms, the surface on which the mouth is located

pearly layer: the smooth, shiny shell layer that is located next to the mantle

pelecypod: mollusk with a shell divided into halves; bivalve

pentaradial symmetry: the body form of echinoderms

prismatic layer: the middle layer of a mollusk shell that is made of calcium carbonate crystals

radula: tongue-like structure in a snail that has a "scraper-like" function

shell: hard, protective covering for the soft body of a mollusk

slug: terrestrial gastropod without a shell

snail: gastropod with a single shell

trochophore: the larval stage in mollusks

true coelom: the body cavity surrounded by mesoderm tissue; it contains many organs

tube foot: a movable suction disk on the ray of most echinoderms

veliger: the second, free-swimming larva into which the trochophore larva develops

visceral hump: structure in an adult mollusk that contains the digestive organs, reproductive organs, excretory glands, and the heart

water-vascular system: the system of tubes that connects to the tube feet in some echinoderms

Name _____ Date _____

Class _____

MOLLUSKS AND ECHINODERMS: BASIC TERMS CROSSWORD PUZZLE

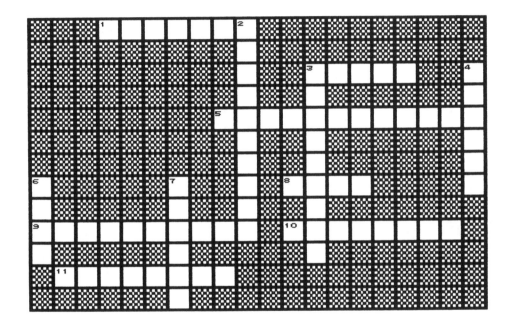

ACROSS

1. Mollusk with two shells
3. Absorb dissolved oxygen from water
5. Interior skeleton
8. A tube _____ is a movable suction disk in many echinoderms
9. Mollusk excretory structure
10. Phylum of snails, slugs, and clams
11. The water- _____ system helps the tube feet work

DOWN

2. Radially symmetrical organism such as starfish and sea urchin
3. "Stomach footed" mollusk such as snail or slug
4. The _____ cavity is formed by the mantle hanging over the body
6. The basic nervous system structure of echinoderms is the nerve _____
7. Rasping tongue

Name _____ Date _____

Class _____

MOLLUSKS AND ECHINODERMS:
ADVANCED TERMS CROSSWORD PUZZLE

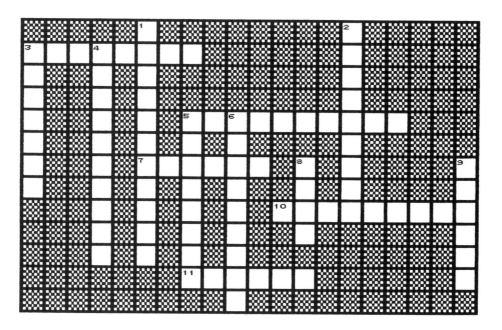

ACROSS

3. The _____ hump contains the heart, digestive organs, reproductive glands, and excretory organs
5. Bilaterally symmetrical starfish larva
7. The _____ layer is smooth and shiny
10. "Stomach-footed" mollusk
11. The _____ surface is away from the mouth

DOWN

1. Larval stage in mollusks
2. Water enters through the _____ siphon
3. The trochophore larva develops into a free-swimming _____
4. Mollusk with reduced or no shell, including squid and octopus
5. The dermal _____ of starfish serves as both a respiratory organ and an excretory organ
6. Bivalve mollusk
8. Surface on which the mouth is located
9. "Rasping tongue" of mollusks

Name _____ Date _____

Class _____

MOLLUSKS AND ECHINODERMS:
VOCABULARY WORDSEARCH

The wordsearch below contains terms related to our study of mollusks and echinoderms. **The words can be found horizontally in either direction, vertically in either direction, and diagonally in either direction.** Clues are given to help you find the words.

```
A B U T R R X J O Q F J G F I J P I Q U
B P Q Z O L J D C D O P O R T S A G O Y
S R F I I U I A R E F E L V T J X Q T V
L R T S B A N B R D R Z V L V B M R N Z
L Y M G L R I P A N S V X T S X T R E O
I G H U V Y O M L N S U K C O B O E R G
G J D B G W O C U T K I G R S M N G R Y
E A P A R T Y L C S C I Y O C A M I U P
R W K T D U E Q S L Z T D J H L P L C H
T A K N L V C R A C P I Q S X O C E N C
N R U E A H H E V T E E X J Z D A V I W
C Z S R R C I M R U B P L A W M B K C A
L O D R O F N G E B C Z H E X B D S E K
S N L U B T O N T E A B U A C I Y N O N
K U Q C A A D D A F X H D R L Y E T F N
U V C X N Y E L W O W Y U F W O P P P M
H K N E N L R M X O V M E O R M P O G N
W G T U A T M M E T S Z T X E X I O D M
N B N R T L L X I E V L A V I B R S D V
D O O M O L L U S C A M U T U X O R R C
```

CLUES

1. Free-swimming larva stage; develops into trochophore
2. Siphon through which water enters the body
3. The _____ surface is away from the mouth
4. Rasping tongue
5. Phylum of snails, clams, and squid
6. Bivalve
7. Water leaves the body through the _____ siphon
8. System of tubes that connect to the tube feet of a starfish; hydrovascular (two words)
9. Absorbs dissolved oxygen from water
10. Examples are starfish, brittle star, and sea urchin
11. The mouth is on the _____ surface
12. Mollusk with no shell or reduced shell and a large head
13. Movable suction disk; part of a hydrovascular system (two words)
14. "Stomach-footed" mollusk
15. Mollusk with two shells

20: ARTHROPODS

VOCABULARY

Basic		Advanced
abdomen	exoskeleton	(all of the basic vocabulary)
antenna	green gland	chelicera
antennule	mandible	Chilicerata
appendage	maxilla	Chilopoda
arachnid	maxilliped	Diplopoda
arthropod	millipede	Isopoda
book lung	molt	Mandibulata
carapace	segmentation	nauplius
centipede	silk	pedipalp
cephalothorax	silk gland	statocyst
cheliped	simple eye	telson
chitin	spider	trilobite
compound eye	spinneret	uropod
crustacean	swimmeret	
	trachea	

DEFINITIONS

abdomen: the posterior segment of an arthropod

antenna: sensory appendage located on the head of many arthropods

antennule: one of a pair of smaller sensory appendages on the head of a crustacean

appendage: projection from the body, such as the antenna of an arthropod

arachnid: one of a class of arthropods with a body divided into two parts, no antennae, simple eyes, and four pairs of legs

arthropod: phylum comprised of animals with jointed, paired legs and hard exoskeleton

book lung: the respiratory organ of a spider

carapace: the hard covering on the back of arthropods such as crabs and lobsters

centipede: member of the class Chilopoda with one pair of legs per body segment

cephalothorax: body region consisting of a jointed head and thorax

chelicera: the first appendage of the spider, used to suck juices from prey

Chilicerata: subphylum of arthropods that includes members of the class Arachnida

cheliped: claw foot in crustaceans

Chilopoda: class of arthropods that includes the centipede

chitin: polysaccharide forming the exoskeleton of arthropods

compound eye: eye composed of numerous lenses

crustacean: member of a class of arthropods that includes lobsters, crabs, shrimp, barnacles, and sowbugs

Diplopoda: class of arthropods that includes the millipede

exoskeleton: hard external skeleton

green gland: excretory organ in crustaceans

Isopoda: order of arthropods that includes the sowbug

mandible: true jaw that works in an up-and-down motion

Mandibulata: subphylum of arthropods that includes the classes Crustacea, Diplopoda, and Chilopoda

maxilla: appendages that aid in chewing with a side-to-side motion

maxilliped: appendage that holds food while it is being chewed

millipede: member of the class Diplopoda with two pairs of legs per body segment

molt: in arthropods, to shed the exoskeleton

nauplius: the larval form of most crustaceans

pedipalp: one of a second pair of appendages of a spider that contains sensory receptors

segmentation: division into units of the bodies of arthropods

silk: a fluid protein that hardens when exposed to air; exuded by the spinnerets

silk gland: organ in spiders that produces fluid that hardens into a silken thread

simple eye: simple light receptor found in many invertebrates

spider: arachnid with pointed appendages, joined head and thorax, and silk glands

spinneret: microscopic tube through which a fluid from the spider's silk glands is passed

statocyst: sac at the base of the antennule of a crayfish that helps it keep its balance

swimmeret: appendage on the abdomen of crustaceans

telson: the posterior segment of the abdomen of certain crustaceans

trachea: air tube in insects and spiders

trilobite: extinct invertebrate thought to be an ancestor of modern arthropods

uropod: the sixth pair of abdominal appendages of a crustacean; developed into a flipper-like structure

ARTHROPODS:
BASIC TERMS CROSSWORD PUZZLE

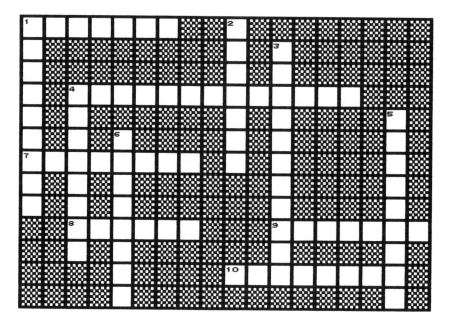

© 1991 by Center for Applied Research in Education

ACROSS

1. Posterior body region
4. Head and thorax combined
7. Member of class that contains spiders, mites, ticks, and scorpions
8. Exoskeleton is made of this
9. Air tube
10. Jaw

DOWN

1. Term for body outgrowth such as an arm or leg
2. Shedding of skin or exoskeleton
3. Exterior skeleton
4. Hard covering on back
5. Abdominal appendage of crayfish
6. Claws of crayfish

Name _____ Date _____

Class _____

ARTHROPODS:
ADVANCED TERMS CROSSWORD PUZZLE

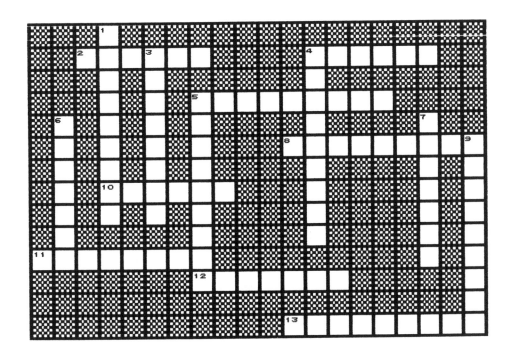

ACROSS

2. Crayfish flipper
4. Exoskeleton is made of this
5. Tube for spider's silk
8. Class containing millipedes
10. Posterior segment of crustacean abdomen
11. Crustacean larva
12. Air tube
13. Has two legs per segment

DOWN

1. Ancient arthropod; found as a fossil
3. Second pair of appendages; acts as sensors for spider
4. First pair of appendages of a spider; "fangs"
5. Helps crayfish keep its balance
6. Group of crustaceans that includes sowbugs and pillbugs
7. Shedding of skin or exoskeleton
9. Anterior crayfish appendage

ARTHROPODS:
VOCABULARY WORDSEARCH

The wordsearch below contains terms related to our study of arthropods. The words can be found horizontally in either direction, vertically in either direction, and diagonally in either direction. Clues are given to help you find the words.

```
J T Z Z W B G E N O I T A T N E M G E S
U P C H U U Z S Q L M Q S L N I T I H C
O E L B I D N A M U I E V M G M G T V Y
J A V E B X U Y K E Y R W U J K Q O Q Z
G G T Q E M P J G X O C X C D G C V M W
A M E Y T W D A E A A R A C H N I D L J
B Q N K R I T C H R M L D G B E S A R Y
M T O N A M N A H O S P O D R V T U H F
J E T E C D O R W H L M P R U M A R V K
F N E M H Q S A Z T F L O G K Q T D A E
T E L O E Z L P W O R I R W E O O I N I
U Q E D A H E A K L X R U F U F C G T V
F W K B N K T C C A H W K G A Y Y L E V
H Y S A T T R E H H H B I Z P Q S E N T
L P O P F O M Q J P G W I A C J T H N W
A Y X U D P E Z H E S K I S O P O D A I
Q H E I P L R B Z C V O Y S N X O N X P
Q Z V D G D Q V F U P X A Z M S T W N L
K M O L T I N G M F Q D I N W M O C F A
D A N A U P L I U S S C R R I K J R Z K
```

CLUES

1. Flattened appendage at posterior end of crayfish
2. Crustacean larva
3. Division of body into sections
4. Exterior skeleton
5. Hard covering on back of crustacean
6. Body region posterior to thorax
7. Posterior segment of abdomen
8. Order of pillbugs and sowbugs
9. Shedding of skin or exoskeleton
10. Forms arthropod exoskeleton
11. Member of the class that includes spiders and mites
12. Balance organ in crayfish
13. Air tube
14. True jaws
15. Combination of head and thorax
16. Organ of touch; appendage of head

VOCABULARY

Basic

amphioxus	instinct
Chordata	intelligent behavior
circulatory system	muscular system
cranium	nervous system
digestive system	notochord
dorsal nerve cord	reflex
endocrine system	reproductive system
endoskeleton	respiratory system
excretory system	skeletal system
gill slit	tunicate
immune system	vertebrae
innate behavior	vertebral column
	vertebrate

Advanced

(all of the basic vocabulary)
Agnatha
appendicular skeleton
Cephalochordata
Chondrichthyes
conditioned response
integumentary system
learned behavior
Osteichthyes
pectoral girdle
pelvic girdle
self-preservation
species preservation
Urochordata

DEFINITIONS

Agnatha: class of vertebrates that includes the jawless fish

amphioxus: free-swimming cephalochordate

appendicular skeleton: the limbs, shoulders, and pelvic girdle in vertebrates

Cephalochordata: subphylum that includes the amphioxus

Chondrichthyes: class of vertebrates that includes cartilagenous fish such as sharks, skates, and rays

Chordata: phylum of the animal kingdom comprised of organisms having a notochord, dorsal nerve cord, and gill slits at some time in their life

circulatory system: organ system that transports blood

conditioned response: a learned response, acquired in reaction to a specific stimulus in a specific way

cranium: part formed by the bones of the skull which encloses the brain

digestive system: the organs for breaking down and absorbing food, includes the mouth, esophagus, stomach and intestine, and accessory organs such as the liver

dorsal nerve cord: hollow tubular structure that lies just above the notochord

endocrine system: the glands that produce and secrete chemical regulators (hormones)

endoskeleton: internal support system made up of bone or cartilage or both

excretory system: the organs, including the paired kidneys, that remove wastes from the body

gill slit: in fish and immature amphibians, structures that act as gills; in other land chordates, these structures develop into various throat structures

immune system: cells and substances in the blood that detect and help destroy pathogens such as bacteria

integumentary system: the outer body covering including skin and special protective outgrowths such as scales, feathers, and hair

innate behavior: behavior that is present from birth, not learned

instinct: unlearned, involuntary action that an animal makes without deliberate decision

intelligent behavior: complex nervous activity involving problem solving, judgment, and decision making

learned behavior: behavior pattern that is not inherited but learned, allowing an animal to alter behavior to cope with change

muscular system: the muscles attached to the skeleton that provide body movement, and the muscles that form the walls of organs such as the heart and stomach

nervous system: the brain, spinal cord, nerves, and special sense organs that together provide sensory perception and voluntary movement

notochord: rod of specialized cells on the dorsal side of lower chordates and present during the embryonic stages of all vertebrates

Osteichthyes: class of vertebrates comprised of the bony fish

pectoral girdle: framework of bones that supports the forelimbs in vertebrates

pelvic girdle: framework of bones that supports the hind legs in vertebrates

reflex: automatic response to a stimulus

reproductive system: the male or female organs that produce gametes

respiratory system: the gills or lungs and related structures used in exchanging gases between the animal and its environment

self-preservation: innate behavior in which an animal will react by escaping or defending itself when in danger

skeletal system: the bones and cartilage that make up the endoskeleton

species preservation: animal instinct that directs reproduction and care of young

tunicate: sessile marine animal belonging to the subphylum Urochordata

Urochordata: subphylum of chordates that includes the tunicate and sea squirts

vertebrae: separate bones comprising the vertebral column

vertebral column: column of bones or cartilagenous structures called vertebrae

vertebrate: member of a subphylum of chordates that has a vertebral column

Name _____   Date _____

Class _____

VERTEBRATES:
BASIC TERMS CROSSWORD PUZZLE

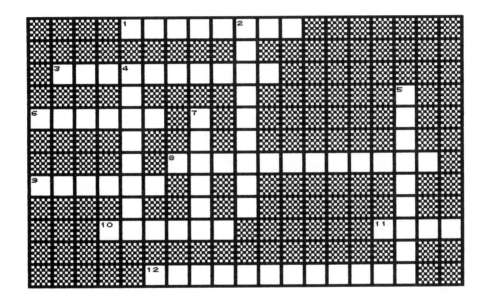

ACROSS

1. Unlearned, involuntary action
3. Has backbone
6. _____ system helps an organism protect itself from diseases
8. Interior skeleton
9. Chordates have a _____ nerve cord
10. Automatic response to stimulus
11. _____ slits are found in the neck area of chordates at some time in their lives
12. Ovaries and testes are part of the _____ system

DOWN

2. The _____ is a rod on the dorsal side, present at least in the embryonic stage
4. Primitive chordate; sea squirt
5. The _____ system produces hormones
7. _____ behavior is present from birth

VERTEBRATES:
ADVANCED TERMS CROSSWORD PUZZLE

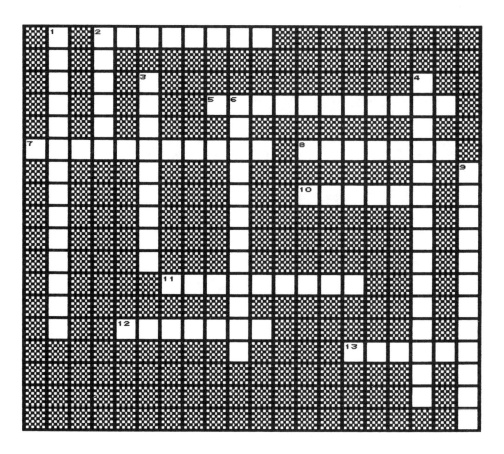

ACROSS

2. The _____ girdle supports the arms or forelimbs
5. _____ responses are a form of learned behavior
7. Subphylum that includes tunicates
8. Class of lamprey and hagfish
10. Chordates have a _____ nerve cord
11. Free-swimming marine animal, subphylum Cephalochordata
12. _____ behavior is not inherited
13. _____ behavior is inherited

DOWN

1. Class containing sharks and rays
2. The posterior appendages such as hind legs are attached to the _____ girdle
3. The _____ is a stiff rod that is replaced by the vertebral column in many chordates
4. Subphylum of amphioxus
6. Class of bony fish
9. Interior skeleton

Name _____ Date _____

Class _____

VERTEBRATES:
VOCABULARY WORDSEARCH

The wordsearch below contains terms related to our study of vertebrates. The words can be found horizontally in either direction, vertically in either direction, and diagonally in either direction. Clues are given to help you find the words.

```
P L J B S H X Q X Z G A X I T Y W L B I
V S E L Q N Y X M E W L G I H S L G C T
U R E J A D I O V O S A G C K G O K Y A
U R P Y H R B E N D O S K E L E T O N E
R E O S H N O H C D Y C R Q R C X J I Y
Z S Q C P T Z T I Z H K I D E N D V H C
V A P W H X H N C B B N X I S E R R A J
L Q B O Q O S C X E Z N L B P F O A G G
H C P K S T R U I P P T C H O P H L O T
G Z T P I T T D W R Y R T G N J C U R T
Z P H N Q Z E N A F D D W Z S B O C U S
A Y C P U P S I S T G N N H E K T I O I
Z T A B Z M F Z C X A J O T O W O D Z J
F M T A X U K T K H V A J H F N N N A R
S G A G N A T H A W T W E L C V U E L P
N C D M Y R P W F G U H B T C V K P P L
N C R R J Q A O E H H K Y U A Z O P T P
D U O Q Y R A T N E M U G E T N I A O O
P W H D R J Z H M P L U F K S M N F U A
P N C J C Y Z O L P Z W X M R T H I F P
```

CLUES

1. Subphylum of tunicates
2. The _____ system includes the skin and hair or other body covering
3. The _____ skeleton includes the arm and leg bones
4. Unlearned, involuntary actions
5. Phylum with notochord, paired gill slits, and dorsal nerve cord
6. The _____ girdle supports the forelimbs
7. Reaction to stimulus
8. Class of chordates without true jaw, scales, or fins; lamprey
9. _____ behavior is unlearned and inherited
10. Class of bony fish
11. Class of sharks and rays
12. Supportive rod on dorsal side of chordates
13. Internal skeleton

107

VOCABULARY

Basic

air bladder
anal fin
artery
atrium
capillary
cartilage
caudal fin
cerebellum
Chondrichthyes
cold-blooded
dorsal fin
external fertilization
gill
gill chamber
lamprey
lateral line

medulla oblongata
milt
mucus
nostril
operculum
optic lobe
Osteichthyes
pectoral fin
pelvic fin
predator
prey
scale
shark
spawn
vein
ventricle

Advanced

(all of the basic vocabulary)
chromatophore
cranial nerve
Cyclostomata
ectothermic
endothermic
integument
olfactory lobe
spinal nerve

DEFINITIONS

air bladder: thin-walled sac that permits fish to remain at a particular depth in the water

anal fin: single fin that grows along the midline on the ventral side of a fish

artery: large, muscular vessel that carries blood away from the heart

atrium: thin-walled upper chamber of the heart that receives blood from the veins

capillary: small, thin-walled blood vessel

cartilage: soft, light, pliable material that makes up the internal skeleton of sharks, skates, and rays

caudal fin: structure that forms a fish's tail

cerebellum: region of the brain concerned with balance and muscular coordination

Chondrichthyes: class of fishes with internal skeletons made of cartilage

chromatophore: pigment-containing structure in the skin of fishes

cold-blooded: describes an animal whose internal body temperature is controlled by its environment

cranial nerve: one of two to twelve pairs of nerves that extend from the brain

Cyclostomata: order of jawless fishes that includes lampreys and hagfishes

dorsal fin: anterior and posterior fins that help fish stay in an upright position while swimming

ectothermic: describes an animal with a body temperature influenced by its environment

endothermic: describes an animal that maintains a constant body temperature by internal regulation of heat loss and gain

external fertilization: fertilization that occurs outside the body of the female

gill: respiratory organ that absorbs dissolved oxygen from water

gill chamber: cavity that encloses the gills

integument: the layer of skin that includes the scales

lamprey: jawless parasitic fish lacking limbs and ribs

lateral line: single row of pitted scales running along each side of a fish that helps the fish to interpret its surroundings

medulla oblongata: structure located at the back of the brain that helps control the activities of the internal organs

milt: sperm-containing discharge of fish

mucus: lubricating solution secreted by mucous glands

nostril: opening that leads to the olfactory lobes

olfactory lobe: region of the brain that regulates smell

operculum: gill cover in fish

optic lobe: region of the brain that regulates sight

Osteichthyes: class of fishes with internal skeletons made of bone

pectoral fin: structure in fishes that is homologous to the front leg of other vertebrates

pelvic fin: structure in fish that is homologous to the hind leg of other vertebrates

predator: an organism that feeds upon another

prey: animal that is killed by a predator

scale: thin, flat disk of bone that grows from pockets in the skin

shark: cartilagenous marine predator

spawn: the mass of eggs discharged by a female fish

spinal nerve: large nerve that connects the spinal column with other parts of the body

vein: vessel that carries blood toward the heart

ventricle: a muscular chamber of the heart

FISHES:
BASIC TERMS CROSSWORD PUZZLE

© 1991 by Center for Applied Research in Education

ACROSS

3. The age of a fish can often be determined by examining the growth rings on a _____
4. An _____ carries blood away from the heart
6. The _____ line is a line of cells along the side of a fish that detects pressure changes in water
10. The heart's _____ pumps blood from the heart to the arteries
13. A shark's skeleton is made of _____
15. The _____ fins are homologous to front legs or arms
16. The air _____ enables the fish to float at a certain level
17. A lubricating solution
18. An example of a member of the class Chondrichthyes is the _____
19. The tail fin is the _____ fin

DOWN

1. The class of the shark and ray
2. Very small, thin-walled blood vessel
3. To discharge gametes into water
5. The _____ fins are homologous to legs
6. Agnathan with circular jawless mouth and rasping teeth
7. Fluid containing sperm
8. Animal that feeds on other animals
9. The _____ fin is a single fin that is on the ventral side
11. The _____ fin is a single fin on top of the fish
12. Water passes over the _____ , which removes oxygen from the water
14. The _____ covers the gills

Name _____ Date _____

Class _____

FISHES:
ADVANCED TERMS CROSSWORD PUZZLE

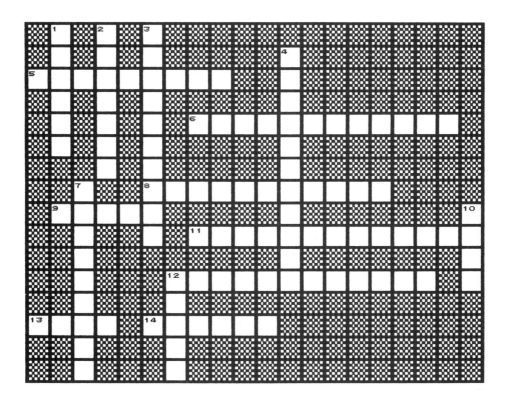

ACROSS

5. The _____ lobe is the center for smell in the brain
6. Order of Agnatha
8. Constant body temperature maintained by organism
9. Release gametes into water
11. Cells contain pigments
12. Class of bony fish
13. Fluid that contains fish's sperm
14. Upper heart chamber that receives blood from veins

DOWN

1. The _____ fins are homologous to arms
2. A _____ nerve is one that extends from the brain to the structures and organs in the head
3. Skin; outer covering
4. Animal with body temperature influenced by environment
7. Gill-opening covering
10. Blood vessel that carries blood to the heart
12. The _____ lobe of the brain registers sight

FISHES:
VOCABULARY WORDSEARCH

The wordsearch below contains terms related to our study of fishes. The words can be found horizontally in either direction, vertically in either direction, and diagonally in either direction. Clues are given to help you find the words.

```
R G S Q C R C N W A P S W H J J K U N N
E N L N Y X M M A H M E N R R L I G I N
R M P E C T O R A L S R V Z E A L W M N
O F E O L W U M G Z C Y I O D T N V U U
H W F M O C E E O O A R S A W E D I U D
P F N U S Y Q T S H L O E C B R J J J C
O L B L T B X A T N E T H H V A M S H I
T V Q U O H Q E E V M C N O U L W E O M
A T Y C M O G G I C J A H N O L G S O R
M F G R A C F A C L I F K D X I N L X E
O X D E T O X L H A S L W R R N F P S H
R D O P A Q R I T M P O Q I H E T X Y T
H Q R O W G X T H P S N Q C X K I Z M O
C C S W R V L R Y R Y M N H M J D V Q T
Y T A I N I O A E E G Z Z T L A D U A C
S P L W M I Z C S Y N C D H D T W M S E
L E E N D O T H E R M I C Y O F M N R O
F R P Y P K E L S I U H E E J L Y B A F
S R V B N Z E H S C S F T S G I L L Y C
D A N A L P E T U L M Y Z S A Y A E B T
```

CLUES

1. The _____ fin is on the ventral side of the fish
2. Members of the class _____ have cartilagenous skeletons
3. Fish remove oxygen from water with this
4. Fluid containing fish sperm
5. The _____ fins are the anterior paired fins
6. Pigment-containing structure
7. The _____ lobe of the brain registers smell
8. Shark skeleton is made of this
9. Maintaining constant body temperature
10. Agnathan
11. Gill cover
12. These cover a fish's body
13. Jawless fishes
14. The tail fin is the _____ fin
15. The _____ fin is on the top of the fish
16. Line of sensory cells on the side of the fish (two words)
17. Class of bony fishes
18. To release eggs into the water
19. Having body temperature influenced by environment

VOCABULARY

Basic		*Advanced*
alimentary canal	kidney	(all of the basic vocabulary)
Amphibia	metamorphosis	Anura
amphibian	newt	Apoda
amplexus	nictitating membrane	coelacanth
cloaca	salamander	conus arteriosis
eustachian tube	toad	glottis
frog	tympanic membrane	maxillary teeth
		mesentery
		torpor
		truncus arteriosis
		Urodela
		vomerine tooth
		winter dormancy

DEFINITIONS

alimentary canal: those organs that compose the food tubes in animals

Amphibia: class of freshwater or terrestrial vertebrates that includes frogs and sala-manders

amphibian: freshwater or terrestrial vertebrate with slimy skin, a three-chambered heart, and which undergoes metamorphosis

amplexus: the clasping of the female frog by the male frog that causes the female to release her eggs

Anura: order of amphibians that includes frogs and toads

Apoda: order of amphibians that are wormlike and lack limbs

cloaca: chamber in certain vertebrates at the end of the digestive, urinary, and reproductive systems

coelacanth: lobe-finned fish whose ancestors may have been related to the ancestors of amphibians

conus arteriosis: the vessel through which blood leaves the amphibian heart

eustachian tube: canal that allows air from the mouth cavity to enter the chamber behind the eardrums

frog: amphibian that lives very close to water and has hind legs well adapted for swimming and jumping

glottis: slitlike opening that leads to the lungs

kidney: organ of excretion in vertebrates that filters nitrogenous wastes from the body

maxillary teeth: small, cone-shaped teeth in the frog's upper jaw that aid in holding prey

mesentery: folded membrane that connects to the intestine and the dorsal body wall

metamorphosis: series of developmental changes in structure that occurs from birth to the completed adult form

newt: very small, land-living salamander

nictitating membrane: in amphibians, third eyelid that helps to keep the eyeball moist when on land

salamander: amphibian with a narrow body, long tail, short legs, and clawless toes

toad: amphibian that lives in loose, moist soil and has hind legs useful for digging and leaping

torpor: state in which body activities stop almost completely

truncus arteriosis: one of two branches into which the conus arteriosis divides after leaving the heart

tympanic membrane: eardrum; in frogs and toads located on the body surface just behind the eye

Urodela: order of amphibians that includes salamanders and newts

vomerine teeth: one of two teeth that projects from bones in the roof of the mouth

winter dormancy: period of inactivity during which an amphibian lies in its burrow in a state of torpor

Name _____ Date _____

Class _____

AMPHIBIANS:
BASIC TERMS CROSSWORD PUZZLE

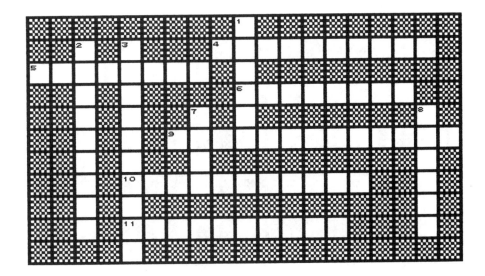

ACROSS

  4. Main "tube" of digestive system; _____ canal
  5. The _____ membrane is the frog's eardrum
  6. Mating by frogs
  9. Change from tadpole to adult
 10. The _____ membrane is a third eyelid
 11. The _____ tube connects the mouth cavity with the middle ear

DOWN

  1. Cavity into which kidney tubes, bladder, and sex organs empty
  2. Term for frogs, toads, newts, and salamanders
  3. Amphibian with a tail as an adult
  7. Kind of salamander
  8. Excretory organ that removes wastes

AMPHIBIANS:
ADVANCED TERMS CROSSWORD PUZZLE

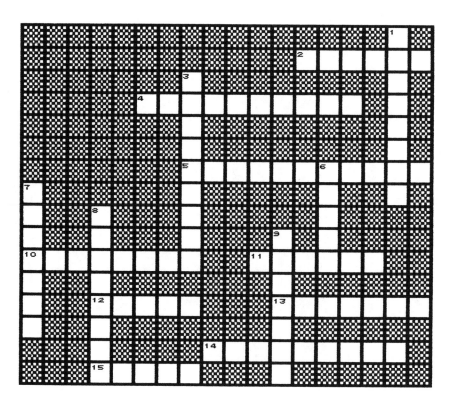

ACROSS

2. State of inactivity
4. Ancient lobe-finned fish
5. The _____ membrane is a third eyelid
10. Winter sleep or _____ is a period of inactivity
11. The chamber that receives fluids from kidneys, bladder, and sex organs
12. The order of legless amphibians
13. The _____ arteriosus is a branch off the conus arteriosis
14. The many _____ teeth surround the frog's upper jaw
15. The _____ arteriosus receives blood from the ventricle

DOWN

1. Two _____ teeth on the roof of the frog's mouth help hold food
3. The membrane that connects interior organs to the body wall
6. The order that includes frogs and toads
7. The order that includes salamanders and newts
8. The _____ membrane is the eardrum
9. The opening to the lungs, at the back of a frog's mouth

Name _____ Date _____

Class _____

AMPHIBIANS:
VOCABULARY WORDSEARCH

The wordsearch below contains terms related to our study of amphibians. The words can be found horizontally in either direction, vertically in either direction, and diagonally in either direction. Clues are given to help you find the words.

```
D V G A N U R A X F D V M K Z H X U N U
I T Y M P A N I C C S S Z B S A D Y Q L
P S Z J I R U R A Y H U P D K K T Z Y G
L N P B O W V G M A D O P A G E M S C N
M V I P Y R A L L I X A M B T E N M N I
J O R I D T P A C D P Q N T T H O H A T
H O O Y M A E L R A P R Z A T B J Q M A
T Y L R B B R E N H Z J M A R O V J R T
I A R E U M W D X Z M O G B N I T C O I
J J R T M I P O Z S R L T A Q K L C D T
U R B N Q X C R O P P N I I F O N E T C
A F P E J Y M U H E K B P M A G I X Q I
R M P S D D Z O F C I Q V C A G D M F N
P V W E L C S A U H H U A B I X E D N Y
P C L M X I Q D P N F U X G W W G H P N
M U Z F S W C M J X S M B Q W P S P L S
T V E B O J A D O F V Q K W U T A Y I E
T I G K C H I E T Q N M O F U Y D B G X
J B O Q K Q S F Y F Q J F X G X B G U J
F I N E X J T T G U Z V W U E S K V H L
```

CLUES

1. Period of inactivity
2. Membrane that holds intestine in place
3. Order of frogs and toads
4. Series of changes from immature form to adult form
5. Order of newts and salamanders
6. The _____ teeth line the frog's upper jaw
7. _____ membrane, the frog's ear

8. Chamber that functions as digestive, excretory, and reproductive duct
9. Body activities stop almost completely
10. Order of wormlike amphibians
11. The _____ membrane is a third eyelid
12. Term for member of the class Amphibia

VOCABULARY

Basic		*Advanced*	
alligator	lizard	(all of the basic	oviparous
amniote egg	molting	vocabulary)	ovoviviparous
caiman	monitor lizard	allantois	parietal eye
carapace	reptile	amnion	plastron
crocodile	Reptilia	Chelonia	Rhynchocephalia
dinosaur	skink	chorion	Squamata
gavial	snake	Crocodylia	tuatara
gecko	terrapin	ecdysis	viviparous
gila monster	tortoise	embryonic membrane	yolk-sac membrane
iguana	toxin	hemotoxin	
internal	turtle	neurotoxin	
fertilization	viper		

DEFINITIONS

allantois: embryonic membrane that aids in respiration and excretion of wastes

alligator: crocodilian reptile that lives in the southern United States and in China

amnion: innermost fetal membrane that forms the sac that encloses the fetus

amniote egg: egg having a protective membrane and a porous shell enclosing the developing embryo

caiman: crocodilian reptile that lives in Central America

carapace: the hard covering on the back of a turtle

Chelonia: order of reptiles that includes turtles and tortoises

chorion: membrane that lines the inside of a reptilian egg

crocodile: crocodilian; large armored reptile with worldwide distribution

Crocodylia: order of reptiles that includes the alligator, crocodile, and caiman

dinosaur: any of an extinct, reptile-like class of vertebrates

ecdysis: the process of molting

embryonic membrane: one of four membranes covering the embryo in an amniote egg

gavial: crocodilian reptile that lives in Southeast Asia

gecko: small, primitive lizard with pads on the undersides of its toes

gila monster: poisonous lizard that lives in the southwestern United States and in Mexico

hemotoxin: poison that breaks down red blood cells and the walls of small blood vessels

iguana: lizard with a series of horny spines on its head and back

internal fertilization: fertilization that occurs inside the body of the female

lizard: long-bodied reptile with tapering tail and usually with four legs

molting: the shedding of a scale layer

monitor lizard: the largest lizard in the world

neurotoxin: poison that affects the nervous system

oviparous: producing offspring from eggs hatched outside the body

ovoviviparous: producing offspring from eggs that hatch inside the mother's body

parietal eye: third eye in the tuatara, it is sensitive to the sun's radiation

plastron: the lower shell of the turtle

reptile: vertebrate with body scales, toes with claws, and amniotic eggs

Reptilia: class of vertebrates that includes turtles, alligators, lizards, and snakes

Rhynchocephalia: order of reptiles that includes the tuatara

skink: lizard with a shiny, cylindrical body and weak, short legs or none

snake: reptile with no limbs, eyelids, or external ear openings

Squamata: order of reptiles that includes lizards and snakes

terrapin: freshwater turtle

tortoise: a turtle that lives on land

toxin: poison produced by a living organism

tuatara: large, spiny reptile of the genus *Sphenodon* that is the only surviving rynchocephalian

turtle: reptile with a shell and horny beak

viper: poisonous snake with long hinged front fangs that fold back along the roof of the mouth when not in use

viviparous: describing animals that bear live young

yolk-sac membrane: embryonic membrane that provides food for the embryo

Name _____ Date _____

Class _____

REPTILES:
BASIC TERMS CROSSWORD PUZZLE

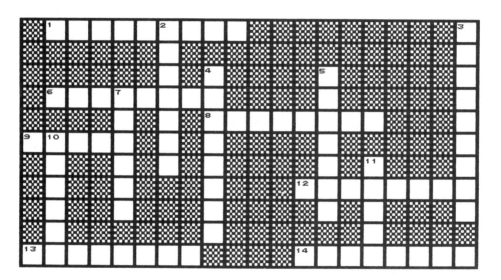

ACROSS

1. Crocodilian of southern United States
6. Land turtle
8. Class that includes snakes and lizards
9. Poisonous snake with long hinged fangs
12. Reptiles have _____ fertilization
13. Turtle shell
14. Ancient extinct reptile

DOWN

2. An _____ egg has a protective membrane and porous shell
3. Large, four-legged, carnivorous reptile
4. Freshwater turtle
5. Shedding of skin or scale layer
7. Reptile with shell
10. Lizard with spines on head and back
11. Lizard with toe pads

Name _____ Date _____

Class _____

REPTILES:
ADVANCED TERMS CROSSWORD PUZZLE

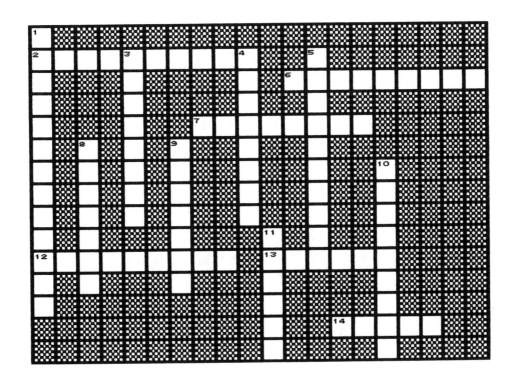

ACROSS

2. Bearing live young
6. Poison that attacks blood
7. Lower shell of turtle
12. Lays eggs
13. Membrane that encloses fetus
14. Lizard with shiny cylindrical body

DOWN

1. Offspring from eggs that hatch inside mother's body
3. The _____ eye is a third eye
4. Order including lizards and snakes
5. Poison that attacks nerves
8. Molting process
9. Membrane that lines inside of egg
10. Embryonic membrane that aids in respiration and excretion
11. Central American crocodilian

REPTILES:
VOCABULARY WORDSEARCH

The wordsearch below contains terms related to our study of reptiles. The words can be found horizontally in either direction, vertically in either direction, and diagonally in either direction. Clues are given to help you find the words.

```
E P T L P Z P F O M Q J W G W I A C J Y
H B W A Y K U D P E Z H M S K D R S Y K
A O I Q H E C A P A R A C V O Y S N S O
N X P Q Z V D G D Q V F U P X A Z M I T
W N L K O H W A A Q S M F Q D I N W S O
C F S D A U T S C Q M B T S C R R I Y J
R Z U Z O V O V I V I P A R O U S S D V
T Z O J Y B V S Q U A M A T A N S V C N
Z C R R N E U R O T O X I N I P H B E S
D T A T A E C A F P Z V R F P A Z K N I
O G P L W E J O R H R S F B U R D T E I
N L I B P H E M O T O X I N B I S T G R
B P V I L U A A F M A Y B V V E U H L W
J A I W F S V P H O L G R W U T O R Q P
T I V Y D H T P W L G Y U T S A R M G P
Q S U M V Q X U C T D Z R M N L A M Y P
B R P D M A G P Z I V X W P B A P R W D
L B J E L G O V U N C X C H B X I B X N
N K J Z N U C O G G I X S E X B V T C H
F I S S A Q I N O M T U M K P G O Y T L
```

CLUES

1. Bearing live young
2. Producing offspring from eggs that hatch inside the mother
3. Poison that attacks the blood
4. Upper turtle shell
5. Order of lizards and snakes

6. Egg laying
7. Process of molting
8. The tuatara has a third eye called a _____ eye
9. Poison that attacks the nerves
10. Shedding of skin

25: BIRDS

VOCABULARY

Basic

air sac	follicle
albumen	gizzard
Aves	incubation
caecum	migration
cloaca	molting
courtship	quill
crop	shell membrane
egg tooth	talon
endothermy	warm-blooded
feather	yolk

Advanced

(all of the basic vocabulary)
altricial bird
Archaeopteryx
barb
barbule
chalaza
contour feather
down feather
precocial bird
quill feather
rachis
syrinx
thecodont
vane

DEFINITIONS

air sac: cavity connected to the lungs that extends into the body cavity

albumen: the egg white that surrounds the egg yolk

altricial bird: bird that is helpless when hatched

Archaeopteryx: extinct feathered dinosaur; ancestor of birds

Aves: class of vertebrates comprised of birds

barb: small ray that composes the vein of a feather

barbule: tiny projection on each barb of a feather

caecum: blind-ended branch of a hollow organ

chalaza: twisted strand of albumen that extends from the egg cell in a bird's egg to the shell membrane

cloaca: a chamber at the end of the digestive, urinary, and reproductive systems

contour feather: feather that rounds out a bird's body and helps to make it streamlined

courtship: behavior pattern that precedes mating

crop: food-storage organ of the alimentary canal

down feather: soft feather of a newly hatched bird

egg tooth: sharp tooth on the end of the bill used by hatchlings to cut through the shell

endothermy: the maintenance of a constant body temperature by internal regulation of heat loss and gain

feather: modified scale found on birds

follicle: indentation in the skin from which a feather grows

gizzard: organ that grinds food in the digestive system of birds

incubation: behavior pattern that keeps eggs warm so that they can mature to hatching

migration: seasonal movement of birds from one environment to another

molting: process of feather replacement

precocial bird: bird that is active as soon as it is hatched

quill: cylinder that originates from the follicle and produces a feather

quill feather: large contour feather that grows in the wing or the tail

rachis: the central axis of the quill of a feather

shell membrane: double lining around the albumen and inside the shell of the egg

syrinx: the song box of a bird

talon: long, sharp claw found on the feet of birds of prey

thecodont: prehistoric lizardlike reptile; possible ancestor of modern birds

vane: the broad flat part of a feather that spreads out from the rachis

warm-blooded: describes an animal that maintains a constant body temperature by internal regulation of heat loss and gain

yolk: special food in the egg that nourishes the developing embryo

Name _____   Date _____

Class _____

BIRDS:
BASIC TERMS CROSSWORD PUZZLE

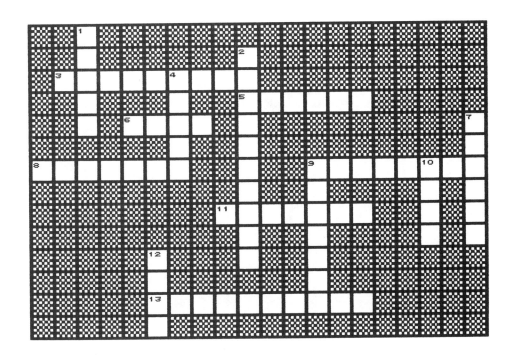

ACROSS

3. Seasonal movement
5. Chamber at end of reproductive, digestive, and urinary systems
6. Embryo's food
8. Egg white
9. Site of feather growth
11. Food-grinding organ
13. Maintenance of constant body temperature

DOWN

1. Cylinder growing from follicle
2. Keeps eggs warm
4. Claw
7. Pouch on hollow organ
9. Modified scale
10. Food-storage area
12. Class of birds

BIRDS:
ADVANCED TERMS CROSSWORD PUZZLE

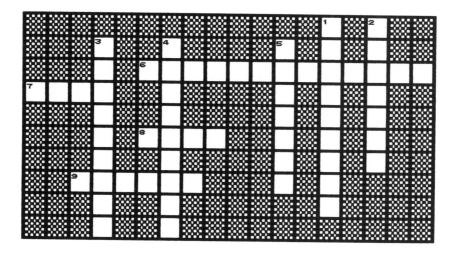

ACROSS

6. Extinct reptile-like bird
7. Broad, flat part of feather
8. Soft feather of newly hatched bird
9. Central axis of feather

DOWN

1. Helpless when hatched
2. Tiny projection from feather's barb
3. Ancient lizardlike reptile
4. Bird that is active as soon as it is hatched
5. Feather that streamlines body

Name _____ Date _____

Class _____

BIRDS:
VOCABULARY WORDSEARCH

The wordsearch below contains terms related to our study of birds. The words can be found horizontally in either direction, vertically in either direction, and diagonally in either direction. Clues are given to help you find the words.

```
D T T E D D N P A O O I L W L Q W M G Q
F M P J N A F C L G D C T D L L V C U F
G D D U C D W O L Z N K Y T S A S C Z J
Q G Q C H P O N D O W N C R Y I L A K B
K G X V P B M T Q Q J L H P Y C Y C I D
P W V Y I N X O H R A C H I S I E C D A
F R W P R X R U Y E V L L F B R G Z F M
R Z E Q B E R R C U R Y H K M T J Z R J
Z M C C M Y T N Z L Z M F T E L Q Z D H
V Z B Y O M A P G U V C Y Q S A E D N Y
B A H N L C U C O H T H E C O D O N T A
N K N J T C I Z D E E R N J W V R Y R N
Q G T E I J Z A S T A V X E L Z E Z S S
X X S O N K C L L E D H H X U E V O Q H
V X J O G R K P P H H N C S A X J K F S
J L E N O F L F F E V C O R T O L I O K
B T L P Z C X W X Q X F Q V A O E V F M
Z T P V C R Q Z C Q B N X Q Y H F E C V
```

CLUES

1. Broad flat part of feather
2. A _____ bird is active as soon as it hatches
3. Extinct feathered dinosaur
4. Losing and replacing feathers
5. Ancient lizardlike reptile
6. Very soft feather type, especially of a young bird
7. An _____ bird is helpless when it hatches
8. Maintenance of a constant body temperature by internal regulation of heat loss and gain
9. Central axis of the quill of a feather
10. A _____ feather rounds out and streamlines a bird's body
11. Food in egg
12. Food-storage organ

26: MAMMALS

VOCABULARY

Basic

anthropoid
carnivore
cerebrum
diaphragm
duckbilled platypus
gestation period
Mammalia
mammary gland

marsupial
mole
omnivore
placenta
placental mammal
primate
rodent
ruminant
thermoregulation

Advanced

(all of the basic vocabulary)
Carnivora
Cetacea
Chiropoda
Edentata
Insectivora
Lagomorpha
Monotremata
Proboscidea
Rodentia
ungulate

DEFINITIONS

anthropoid: a member of a group of mammals that generally has an upright posture

Carnivora: an order of flesh-eating mammals that includes bears, weasels, minks, otters, and lions

carnivore: meat-eating organism

cerebrum: the largest region of the [human] brain; the seat of emotions, senses, intelligence, and voluntary muscular activity

Cetacea: order of mammals that includes whales, porpoises, and dolphins

Chiropoda: order of mammals that includes bats

diaphragm: muscle that separates the abdominal and thoracic cavities and is involved in the process of breathing

duckbilled platypus: egg-laying mammal with a bill used to probe for worms and grubs

Edentata: order of toothless mammals that includes armadillos, sloths, and anteaters

gestation period: the period of development in the uterus in mammals

Insectivora: order of insect-eating mammals that includes moles and shrews

Lagomorpha: order of rodentlike mammals that includes rabbits and hares

Mammalia: class of vertebrates that includes animals with body hair, are warm-blooded, and have mammary glands

mammary gland: gland in female mammals that secretes milk

marsupial: mammal that is born immature and completes its development in the pouch of the mother

mole: small underground mammal that eats insects

Monotremata: the order of egg-laying mammals

omnivore: animal that eats both plants and meat

placenta: oval, spongy structure in the uterus that transports substances between the mother and the developing young

placental mammal: a mammal whose entire prenatal development takes place in the uterus of the mother

Primates: order of upright mammals with the most highly developed brain and nervous system, includes lemurs, gibbons, orangutans, gorillas, and humans

Proboscidea: order of trunk-nosed mammals such as the elephant and the fossil mammoth

rodent: mammal of the order Rodentia

Rodentia: order of gnawing mammals such as squirrels, mice, and rats

ruminant: hoofed mammal that has a complex, three- or four-chambered stomach

thermoregulation: the ability to internally regulate body temperature

ungulate: hoofed mammal

MAMMALS:
BASIC TERMS CROSSWORD PUZZLE

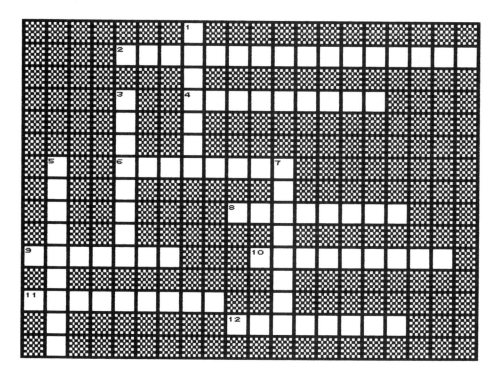

ACROSS

2. The ability to internally regulate body temperature
4. A pouched mammal
6. The largest area of the brain; the site of voluntary nervous activity
8. A hoofed mammal
9. The gland that secretes milk
10. Meat eater
11. The breathing muscle; it separates the thorax from the abdomen
12. An egg-laying mammal

DOWN

1. Order of upright mammals with the most highly developed brain and nervous system
3. The structure that transports material between the mother and the fetus
5. The period of development in the uterus
7. Class of vertebrates that has body hair, are warm-blooded, and nurse young with mammary glands

Name _____ Date _____

Class _____

MAMMALS:
ADVANCED TERMS CROSSWORD PUZZLE

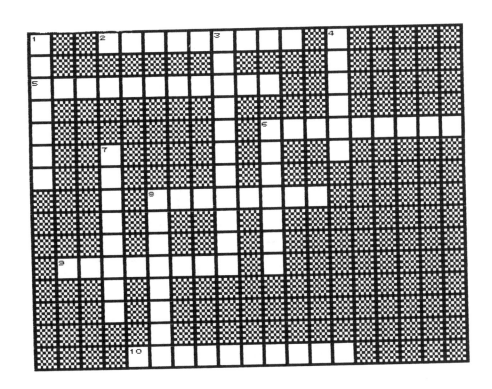

ACROSS

  2. Pouched mammal
  5. Order of insectivorous mammals, including shrews and moles
  6. Order of meat-eating mammals, including cats, dogs, and bears
  8. Hoofed mammal
  9. Structure that nourishes the fetus
 10. Order that includes rabbits and hares

DOWN

  1. Erect-walking mammal, like an ape or a human
  3. Order that includes the elephant
  4. Mammal such as the squirrel and the mouse
  6. Order that includes whales and porpoises
  7. Order of toothless mammals, including armadillos and sloths
  8. Order that includes squirrels, rats, and mice

MAMMALS:
VOCABULARY WORDSEARCH

The wordsearch below contains terms related to our study of mammals. The words can be found horizontally in either direction, vertically in either direction, and diagonally in either direction. Clues are given to help you find the words.

```
C T O Q J E U J U Z D G K P F K D S E S
S J M N F L P R U V D H V G A X Z V F M
V I N V D A R O V I N R A C B Z Q C I S
B Y I J M I I A C F P Y R M Z T N N X U
B N V V P P M P T R U R D H E Q S V S P
K Y O O I U A N X Y N U J N C E L T H Y
Y G R A O S T V W O Y P B P C V S A K T
U J E T L R E M I C J B L T T A E Q F A
O E T J F A O T B A E A I Z I D V V V L
U O H W R M A L H P C V H T I T D X T P
H N N E C T X P B E O M N C I P G U N X
G Q N E S Q R V N R I E S H P S K N A P
A U N E C O Y T A V D O G L D T T G N Y
U M G W M U A T E O B O F X D E E U I H
I R N O E S L B R O K U P C C L V L M B
K H G I Y G Z S R I J U O L V R Q A U P
Y A Z U J R P P S Q A E D W K C M T R B
L I T Q G R H Z X N Y J O W K U P E I P
G V X B M Z R R W T Y A E C A T E C S A
F F A Z T R N Q W Y M U H E I W E I T H
```

CLUES

1. Hoofed mammal
2. Order of rabbits and hares
3. Order of meat-eaters such as cats, dogs, and bears
4. Structure that transports materials between the mother and the fetus
5. The _____ period is the period of development in the uterus
6. Order of gnawing mammals such as squirrels, mice, and rats
7. Order of insect-eating mammals such as moles and shrews

8. Hoofed mammal with specialized digestive system
9. Pouched mammal
10. Animal that eats both plants and meat
11. Order of trunk-nosed mammals such as the mammoth and the elephant
12. Order of whales and porpoises
13. Mammal such as the monkey, ape, lemur, and human
14. The duckbilled _____ is an egg-laying mammal

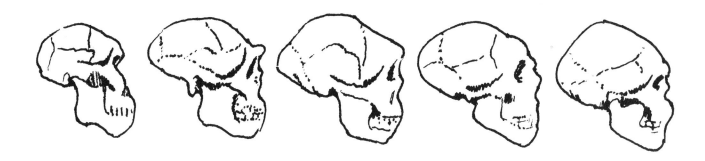

VOCABULARY

Basic

anthropology
baboon
bipedal locomotion
chimpanzee
Cro-Magnon
Jane Goodall
half-life
Homo sapiens

human culture
humanoid
Mary Leakey
Louis Leakey
Neanderthal Man
opposable thumb
Primates
radiocarbon dating

Advanced

(all of the basic vocabulary)
anthropomorphism
Australopithecus
Homo erectus
potassium-argon dating
sociobiology

DEFINITIONS

anthropology: the study of primates, fossil evidence of prehistoric human culture, and all present human cultures

anthropomorphism: giving human traits or characteristics to animals or things that are not human

Australopithecus: hominid genus whose fossil remains were found in Africa

baboon: genus of large ape of Africa and Asia

bipedal locomotion: the ability to walk on two hind limbs

chimpanzee: primate closely related to humans

Cro-Magnon: primitive extinct humans that lived in Europe about 50,000 years ago

Jane Goodall: scientist who has spent most of her adult life studying the behavior of chimpanzees

half-life: the amount of time for one half of a given amount of a radioactive isotope to reach its stable form

Homo erectus: hominid that lived from about one million to about 150,000 years ago

Homo sapiens: the species that includes all the people living today

human culture: system of traditions, language, and learning passed on from one generation of humans to the next

hominid: a humanlike organism that is extinct

Mary Leakey: scientist who discovered a fossil hominid dated to about 3,600,000 years ago

Louis Leakey: an anthropologist who, along with his wife Mary, discovered many fossil hominid remains

Neanderthal man: extinct *Homo sapiens* that lived from about 150,000 years to about 35,000 years ago

opposable thumb: thumb that can be bent to a position opposite the fingers

potassium-argon dating: radioactive dating method used to determine the age of rock samples

Primates: order of mammals that includes the lemurs, tarsiers, monkeys, apes, and humans

radiocarbon dating: process of radioactive carbon measurement used to indicate the age of a fossil

sociobiology: the study of the effect of biological factors on social systems and societies

Name _____ Date _____

Class _____

HUMANS:
BASIC TERMS CROSSWORD PUZZLE

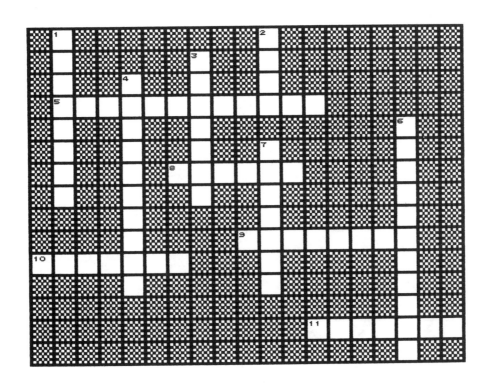

ACROSS

5. Study of primates, especially man
8. Cro-_____ man lived in Europe about 50,000 years ago
9. The amount of time for one half of a sample of radioactive isotope to decay
10. All people living today are in the species *Homo* _____
11. Member of the same order of mammals as man, gorillas, monkeys, and lemurs

DOWN

1. Humanlike
2. Genus for humans
3. Ability to walk on two hind limbs
4. Primate; closely related to man
6. _____ man lived about 100,000 years ago
7. Jane _____ has done extensive studies of chimpanzee behavior

HUMANS:
ADVANCED TERMS CROSSWORD PUZZLE

ACROSS

3. Giving human traits to nonhuman things
4. Study of social behavior of organisms
6. All modern people are in the species *Homo* _____
7. _____ locomotion is the ability to walk on two appendages
10. *Homo* _____ lived about one million years ago
11. _____ man lived about 100,000 years ago

DOWN

1. Humanlike
2. Cro-_____ man lived about 50,000 years ago
3. Fossil remains found in Africa; lived over 1.5 million years ago
5. Time for one half of a sample of a radioactive isotope to decay
8. _____-argon dating is useful in dating fossils more than 50,000 years old
9. _____ dating is useful for fossils less than 50,000 years old

Name _____ Date _____

                                               Class _____

HUMANS:
VOCABULARY WORDSEARCH

The wordsearch below contains terms related to our study of humans. The words can be found horizontally in either direction, vertically in either direction, and diagonally in either direction. Clues are given to help you find the words.

```
N L C W D E W S C C Y X K J V N S F W X
X Y V W Z A Y J O Y X Z Z L S S Q P G T
Y R C X W Q Q O Z S V J U N O J R K S A
B A V J T Y C G E L S M N C A I Q C Q Z
V D J L D B G C H S M U I U M C X Q F L
K I X U A H Y E H K V O T A L X U G X A
Z O Z W B D P Y D I B I T C C U M D E H
S C S W L E E Z Q I M E Y S E B N J L T
Y A G M Z P T P O Q G P U P A R F R U R
A R V U D F Q L I Z E K A D W A E Z E E
X B W I T V O H D B W F Q N H Y G T B D
F O P S V G X H G C B A E G Z P G P Q N
M N R S Y P F B V W L C Q N M E O K L A
F T U A H O M O S A P I E N S P E I Z E
B P P T G B L R Y J W V T O C X I L L N
R F U O B W N O H P Z R O E T O K T D J
M A L P Q M E O O B D Y M L F E A T V T
K L Z B Q V A N T H R O P O L O G Y J N
S D I O N A M U H M A T K L F X Q O B B
R P C J A Y Q O A V Z U W I I U P M B K
```

CLUES

1. Study of social behavior
2. _____ dating can be used to find the age of fossils less than 50,000 years old
3. Humanlike
4. Walking on two limbs
5. _____-argon dating can be used to find the age of fossils more than 50,000 years old
6. Mammal order that includes apes and man
7. Scientific name for modern man (two words)
8. Study of primates and human culture
9. *Homo* _____ lived about 1,000,000 years ago
10. _____ man lived about 100,000 years ago
11. The _____ is a primate that is very closely related to humans

137

VOCABULARY

Basic

abdominal cavity
angular joint
appendicular skeleton
atrophy
axial skeleton
ball-and-socket joint
bone
cardiac muscle
cartilage
clavicle
compact bone
connective tissue
cranial cavity
diaphragm
epithelial tissue
extensor
flexor
gliding joint
hinge joint
immovable joint
insertion
involuntary muscle
joint
ligament
motor unit
muscle tissue
nervous tissue

origin
ossification
pelvis
pericardial cavity
pivot joint
red marrow
rib
sacrum
skeletal muscle
skeletal system
skull
smooth muscle
spinal cavity
spine
spongy bone
sternum
striated muscle
suture
tendon
thoracic cavity
tone
vertebra
vertebral column
voluntary muscle
visceral muscle
yellow marrow

Advanced

(all of the basic vocabulary)
abductor
actin filament
adductor
cervical vertebra
Haversian canal
lumbar vertebra
myofibril
myofilament
osteocyte
periosteum
thoracic vertebra

© 1991 by Center for Applied Research in Education

DEFINITIONS

abdominal cavity: the body cavity that is separated from the thoracic cavity by the diaphragm

abductor: voluntary muscle that moves a limb away from the body

actin filament: thin filament that makes up part of the myofibril

adductor: voluntary muscle that moves a limb toward the body

angular joint: joint that allows movement in two directions, such as a wrist or an ankle

appendicular skeleton: bony structure that makes up the limbs and shoulder and pelvic girdles

atrophy: decrease in the amount of tissue or size of an organ often caused by lack of use

axial skeleton: the bones of the skull, vertebral column, sacrum, ribs, and sternum

ball-and-socket joint: type of joint found in the shoulder and hip

bone: a connective tissue forming the skeleton of vertebrates

cardiac muscle: muscle that makes up the heart wall

cartilage: a hard but flexible tissue in the skeleton of vertebrates

cervicle vertebra: one of the bones in the neck region of the backbone

clavicle: the collarbone

compact bone: forms the hard outer layer of the long bones and has fewer cavities than spongy bone

connective tissue: tissue that binds together and supports other structures

cranial cavity: cavity made up of a number of fused bones that surround and protect the brain

diaphragm: muscle that separates the abdominal and thoracic cavities

epithelial tissue: tissue that covers the body surfaces externally and internally and secretes or absorbs material

extensor: voluntary muscle that straightens joints

flexor: voluntary muscle that bends joints

gliding joint: joint such as found in the bones of the hand and the foot

Haversian canal: channel in bone containing blood vessels

hinge joint: joint that allows free movement in a single plane such as an elbow or a knee

immovable joint: joint that is present when bones meet and no movement occurs

insertion: the attachment of muscles to the movable part

involuntary muscle: muscle that cannot be consciously controlled

joint: the area where two bones meet

ligament: tough strand of connective tissue that holds movable joints in position

lumbar vertebra: one of the vertebrae in the abdominal region of the backbone

motor unit: combination of nerve cells and muscle fibers

muscle tissue: tissue that functions in movement and in other body processes

myofibril: bundle of actin and myosin filaments

myofilament: the thick filament that makes up part of the myofibril

nervous tissue: tissue that functions in communication and coordination of body movements

origin: point of attachment of muscles to the stationary part of a joint

ossification: the process of bone formation

osteocyte: bone cell

pelvis: the cavity at the lower end of the trunk surrounded by the pelvic girdle

pericardial cavity: sac enclosing the heart

periosteum: tough fibrous membrane that covers the bones

pivot joint: joint that allows rotary movement such as in the head and neck

red marrow: soft tissue found inside bones that forms the red blood cells and most of the white blood cells

rib: flattened, curved bone in the chest region

sacrum: bone in the lower region of the backbone consisting of fused vertebrae

skeletal muscle: voluntary muscle attached to bones to move parts of the skeleton

skeletal system: human body system that provides support and form

skull: the most anterior part of the skeleton consisting of the cranium and the facial bones

smooth muscle: type of muscle found in the walls of the blood vessels, the stomach and intestines, and other internal organs

spinal cavity: the cavity that is enclosed by the vertebral column

spongy bone: porous type of bone found in the long bones

sternum: the breastbone

striated muscle: voluntary muscle also known as skeletal muscle

suture: the irregular seams of bones that are coming together

tendon: connective tissue that joins muscle to bone

thoracic cavity: the cavity that contains the lungs, trachea, heart, and esophagus

thoracic vertebra: one of the vertebrae in the chest region of the spine

tone: healthy, normal condition of muscles

vertebra: one of the bones of the spine

vertebral column: flexible jointed column made of bones called vertebrae; spine

voluntary muscle: muscle that is consciously controlled

visceral muscle: involuntary muscle; smooth muscle

yellow marrow: soft tissue found in bones that is made up mostly of fat cells

HUMAN BODY PLAN—SKELETAL AND MUSCULAR SYSTEMS: BASIC TERMS CROSSWORD PUZZLE

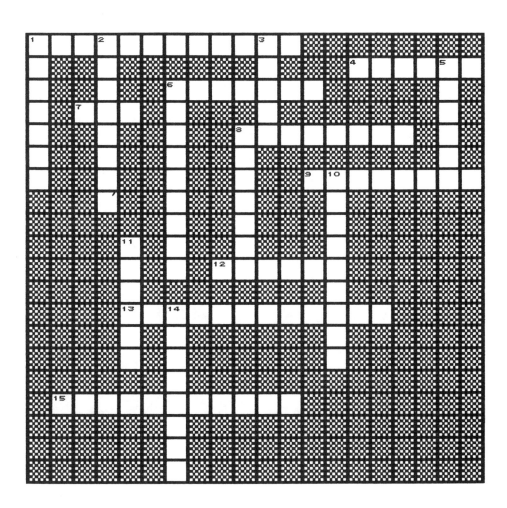

ACROSS

1. The _____ skeleton includes the arm and leg bones
4. _____ marrow is found in the center of long bones and is primarily fat cells
6. _____ muscle is heart muscle
7. _____ marrow is found in flat bones and forms red corpuscles
8. Collarbone
9. Holds bones in position
12. _____ muscles bend a joint
13. Formation of bone
15. The _____ cavity encloses the heart

DOWN

1. To degenerate from lack of use
2. _____ muscles straighten joints
3. The _____ skeleton includes the skull and backbone
5. Point of attachment of a muscle to the immovable part of a joint
6. Tough, flexible substance that is mostly replaced by bone in adult vertebrates
8. The _____ cavity is formed by the skull
10. Attachment of muscle to movable part of a joint
11. _____ muscles show no striations; are also called visceral muscles
14. _____ muscles are generally voluntary and are skeletal

HUMAN BODY PLAN—SKELETAL AND MUSCULAR SYSTEMS: ADVANCED TERMS CROSSWORD PUZZLE

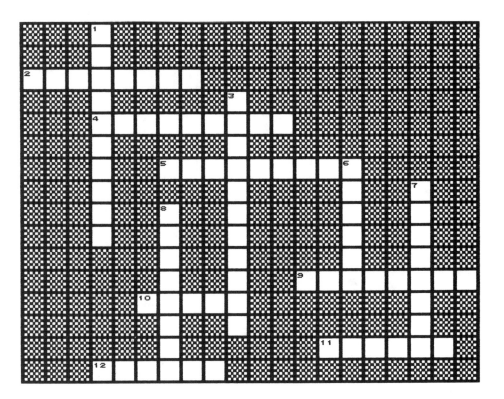

ACROSS

2. The _____ vertebrae are in the upper part of the trunk
4. Bone cell
5. Fine thread of muscle fiber
9. _____ vertebrae are in the neck
10. _____ filaments are the thinner of the two kinds of filaments in myofibrils
11. Seam in bone such as the skull
12. Fused vertebrae in pelvic area

DOWN

1. Surrounds bones, nourishes them, and provides attachment point for muscle
3. Thicker of the two filaments in myofibrils
6. The _____ vertebrae are in the lower back
7. _____ muscles move limbs toward the body
8. _____ muscles move limbs away from the body

Name _____ Date _____

Class _____

HUMAN BODY PLAN—SKELETAL AND MUSCULAR SYSTEMS: VOCABULARY WORDSEARCH

The wordsearch below contains terms related to our study of the skeletal and muscular systems. The words can be found horizontally in either direction, vertically in either direction, and diagonally in either direction. Clues are given to help you find the words.

```
J I K H N P X Q P R Z P V N H N V X Z T
B I J E G E R L W S F M Z D G E R V D T
Z L X M O L Q M C V I P G K P B D D I Y
M I R S L V C Y S Z A L G J Q I Y K G L
E X S S A I D O T I Y Z O F A Y I Z U P
D L W J I S Y F E D F Q Z P B J C Q X X
L R A N X K M I R T M F H Z H S I R G X
I M K C A K X B N O M R T Y P R S D Q I
G N N I I T I R U L A G A N Z O G A P I
A O B A M V T I M G L O R Q V S O D W Q
M D P B Q F R L M W S B X Q Q N W D J X
E N S D N D A E B H C I Q E L E I U O Y
N E T O Y U W N C Y S M V E T T M C Y V
T T D M E F R E D M A R R O W X X T Q Y
J F S I N T R A A L O B C Q Q E M O U F
U X B N F H N N K P Z P B G B U G R O W
O K M A A V P E R I O S T E U M A M K I
O K T L Y A L C H N Y D X A G P K N W N
Y C D Z N E D Y E E V N A N O T K W A W
K P H R A T R G X A Q X M Q X I Z Z N N
```

CLUES

1. Membrane that covers bones
2. Muscle that moves a limb toward the body
3. Breastbone
4. Connective tissue that holds movable joints in position
5. The _____ skeleton is made up of the skull, vertebral column, ribs, and sternum
6. Bundle of actin and myosin filaments
7. The _____ cavity contains the digestive system
8. Soft tissue inside bones; forms red blood cells (two words)
9. An _____ muscle straightens joints
10. A _____ vertebra is in the neck region
11. Joins muscle to bone
12. Surrounded by pelvic girdle; region of lower limb attachment
13. Breathing muscle; separates thoracic and abdominal cavities

VOCABULARY

Basic

absorption	mechanical process
assimilation	mineral
bile	molar
calorie	mucus
canine tooth	neck
carbohydrate	nutrition
chemical process	organic nutrient
colon	pancreas
common bile duct	pepsin
crown	pharynx
digestion	protein
emulsion	reabsorption
enamel	rectum
epiglottis	root
esophagus	saliva
fat	salivary amylase
feces	salivary gland
gallbladder	small intestine
gastric juice	stomach
gum	taste bud
incisor	tooth
inorganic nutrient	tongue
jaw	villi
large intestine	vitamin
liver	water

Advanced

(all of the basic vocabulary)
ascending colon
cementum
descending colon
duodenum
ileum
jejunum
lactase
lipase
maltase
palate
pancreatic amylase
peptidase
periodontal membrane
pulp cavity
pyloric valve
soft palate
sucrase
transverse colon

DEFINITIONS

absorption: any process in which soluble substances pass through a membrane

ascending colon: the part of the colon that passes up the right side of the abdomen

assimilation: process in which digested and absorbed food is used by body cells for growth and maintenance

bile: enzyme that breaks down fat globules into small droplets

calorie: the amount of heat needed to raise the temperature of one gram of water one degree centigrade

canine tooth: large, cone-shaped tooth used for tearing food

carbohydrate: organic compound made of carbon, hydrogen, and oxygen; includes sugars, starches, and cellulose

cementum: the covering of the root of a tooth

chemical process: process in which enzymes complete digestion

colon: the large intestine

common bile duct: a Y-shaped duct that carries bile from the gallbladder to the duodenum

crown: the top part of the tooth, outside of the gum

descending colon: the part of the colon that passes down the left side of the abdomen

digestion: the process in which food is broken down into molecules and absorbed

duodenum: the first section of the small intestine

emulsion: milky substance in which droplets of fat are suspended

enamel: hard white covering on the crown of a tooth

epiglottis: flap of cartilage located at the upper end of the trachea

esophagus: the food tube that connects the mouth with the stomach

fat: organic compound made up of glycerol and fatty acids

feces: waste product containing undigested cellulose, fat, and connective tissue

gallbladder: sac in which bile is stored and concentrated

gastric juice: secretion of the stomach

gum: tissue that supports the teeth

ileum: the last section of the small intestine, joins the large intestine

incisor: flat front tooth used for cutting food

inorganic nutrient: nutrient that does not contain carbon, such as water or minerals

jaw: the hinged joint that enables chewing

jejunum: the middle section of the small intestine

lactase: digestive enzyme that converts lactose to glucose or galactose

large intestine: the digestive organ that reabsorbs water from undigested food; colon

lipase: digestive enzyme that splits fats into fatty acids and glycerol

liver: digestive organ that secretes bile

maltase: digestive enzyme that splits maltose into glucose molecules

mechanical process: the first part of digestion in which food is broken up by the teeth and churned by the muscular movements of the stomach

mineral: inorganic nutrient such as sodium or calcium

molar: flat tooth that grinds and crushes food

mucus: lubricating solution secreted by mucous glands

neck: the narrow part of the tooth at the gum line

nutrition: includes all the processes by which an animal takes in, chemically changes, absorbs, and uses food

organic nutrient: nutrient that contains carbon such as carbohydrates, fats, and proteins

palate: the roof of the chewing area and the part of the mouth that extends to the top of the throat

pancreas: the digestive organ that produces insulin and pancreatic fluid

pancreatic amylase: digestive enzyme that changes starch into maltose

pepsin: digestive enzyme that splits complex protein molecules into shorter chains

peptidase: digestive enzyme that completes protein digestion by breaking the remaining peptide bonds in proteins

periodontal membrane: fibrous tissue that anchors the tooth root firmly in the jaw socket

pharynx: the throat cavity

protein: organic nutrient made up of amino acids

pulp cavity: the center of the tooth that contains blood vessels and nerve fibers

pyloric valve: valve located between the stomach and the small intestine

reabsorption: the process in which water that has already gone through the digestive process is taken back into the body through the walls of the large intestine

rectum: the second part of the large intestine

root: the part of the tooth beneath the surface

saliva: fluid substance secreted by the salivary gland

salivary amylase: digestive enzyme found in saliva that changes starch into maltose

salivary gland: one of three pairs of glands in the mouth that produces saliva

small intestine: the digestive organ in which most of the absorption of nutrients occurs

soft palate: the part of the mouth made of folded membranes that extend from the hard palate to the top of the throat

stomach: the digestive organ that receives ingested food and prepares it for digestion

sucrase: digestive enzyme that splits sucrose, changing it to the simple sugars glucose and fructose

taste bud: oval structure on the tongue surface that has nerve endings at its base

tooth: a hard structure in the upper and lower jaws used for chewing and biting

tongue: a muscular organ in the mouth that functions in tasting, chewing, and swallowing food

transverse colon: the part of the colon that passes from the right to the left side just below the duodenum

villi: fingerlike projections in the small intestine that function in the process of absorption

vitamin: organic substance necessary in small amounts for body growth and activity

water: a common compound, H_2O, that is extremely important to living things because of its ability to dissolve other substances

Name _____ Date _____

Class _____

HUMAN BODY PLAN—DIGESTIVE SYSTEM: BASIC TERMS CROSSWORD PUZZLE

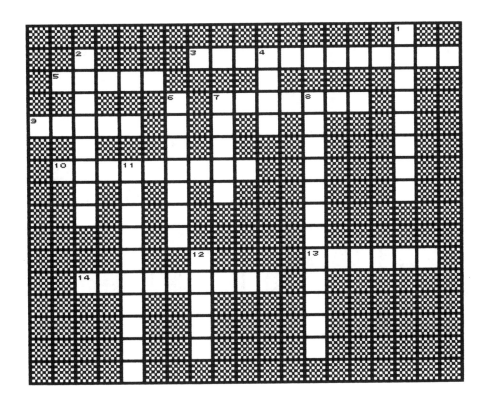

ACROSS

3. Made of carbon, hydrogen, and oxygen; sugars, starch, and cellulose
5. Most protein and sugar digestion occurs in the _____ intestine
7. Unit of energy
9. Projections into the small intestine
10. Breakdown of food into small water-soluble particles
13. Enzyme secreted by the stomach
14. Tube from the mouth to the stomach

DOWN

1. Produces insulin and some digestive fluids
2. Fluid containing suspended food
4. Breaks fat globules into small droplets, forming an emulsion
6. _____ juice or fluid is found in the stomach
7. Large intestine
8. _____ of water occurs in the large intestine
11. Flap of cartilage at upper end of the trachea
12. The _____ intestine receives water masses of undigested food from the small intestine

HUMAN BODY PLAN—DIGESTIVE SYSTEM: ADVANCED TERMS CROSSWORD PUZZLE

ACROSS

1. The _____ colon goes across the abdomen
3. Material moves from the stomach into the duodenum through the _____ valve
6. Chemical that breaks down peptides into amino acids
7. First section of small intestine
9. _____ nutrients include sugars, starches, and proteins
10. Breaks down fats
12. Chewing tooth
13. The _____ colon passes up the right side of the abdomen
14. Second section of small intestine

DOWN

2. Breaks sucrose into glucose and fructose
3. The _____ membrane anchors the tooth to the jaw
4. The gall _____ stores liver bile
5. A simple sugar
8. Necessary organic substance
10. Breaks down lactose
11. Lower part of small intestine

Name _____ Date _____

Class _____

HUMAN BODY PLAN—DIGESTIVE SYSTEM: VOCABULARY WORDSEARCH

The wordsearch below contains terms related to our study of the digestive system. The words can be found horizontally in either direction, vertically in either direction, and diagonally in either direction. Clues are given to help you find the words.

```
G S D E P P Z B H P D E H H N T I B J E
L T V D Q E W N F E D C R N L N I Q H I
A P U V G C G N I D N E C S E D Q E L F
X F F P T J L N F L N Q Q L R V H E Z S
S X C B F Y W I E I V Z Y M O B E W P F
Q R B E X W A Q U P R Q R Y S S Z Y S T
D E S A T C A L S A S X O E R T L Z J N
W J Y Z I D R P O S I M R E J O K K O K
Q V I T A M I N M E I T V V R P U S N M
N R G A H V C L B J K S C I P I D P O L
P K G V O M Q M E T N A C U D X N V H D
W S Y B F D E A L A L R F F K K P S W X
P A N C R E A S R O I R P Y A S I B G R
C D T K C W J T R I F E F J E L I M O R
T R R G F G U I A A B O Y G E P Q U B C
S B Z I R J E Z C U Y H E U Y P A E L G
T B U N U Z D O F A P Z M T O H F P Z V
X P X W F Z H I Z I Z E T A L A P Y X J
T V P M C Q L U E T A R D Y H O B R A C
J O S R K O R C I R T S A G H V N X A V
```

CLUES

1. The _____ colon crosses the abdomen just below the duodenum
2. Enzyme that breaks down fats
3. The _____ colon passes down the left side of the abdomen
4. _____ juice or fluid is secreted by the stomach
5. The _____ valve separates the stomach from the small intestine
6. Enzyme that breaks down lactose
7. Organic nutrient that is needed in very small amounts
8. Organic compound made of carbon, hydrogen, oxygen; examples are sugars and starches
9. Roof of the mouth
10. Last section of the small intestine before it joins the large intestine
11. Organ that produces insulin
12. Amount of heat

VOCABULARY

Basic		*Advanced*
adenoid	lymph node	(all of the basic vocabulary)
antigen	lymphatic system	agglutinate
aorta	medulla	aortic arch
arteriole	nephron	atrioventricular node
artery	perspiration	atrioventricular valve
atrium	plasma	Bowman's capsule
bladder	platelet	coronary circulation
blood	pulmonary artery	diastole
blood solid	pulmonary vein	fibrin
capillary	pulse	fibrinogen
cardiac muscle	receptor	glomerulus
circulation	red blood cell	inferior vena cava
cortex	renal artery	pericardium
dermis	renal vein	portal circulation
epidermis	Rh factor	pulmonary circulation
erythrocyte	skin	renal circulation
excretion	tonsil	semilunar valve
excretory system	ureter	septum
heart	urethra	serum albumin
hemoglobin	urine	serum globulin
homeostasis	vein	sinoatrial node
kidney	ventricle	subcutaneous tissue
leucocyte	venule	superior vena cava
lymph	white blood cell	systemic circulation
		systole
		thrombin

DEFINITIONS

adenoid: mass of lymphatic tissue found in the throat

agglutinate: to clump together

antigen: substance that stimulates the formation of antibodies in the body

aorta: the large artery that leads from the heart to the rest of the body

aortic arch: large curved part of the aorta directly above the heart

arteriole: smaller artery that penetrates into the tissues

artery: blood vessel that carries blood away from the heart

atrioventricular node: small mass of specialized heart muscle tissue located in the right atrium that controls the contractions of the ventricles

atrioventricular valve: a heart valve found between the atrium and the ventricle

atrium: the thin-walled upper chamber of the heart

bladder: the organ in which urine collects after being produced in the kidneys

blood: liquid connective tissue that carries organic and inorganic nutrients and oxygen to the cells, and carbon dioxide and nitrogenous wastes away from the cells

blood solid: one of three different types of cells suspended in blood plasma

Bowman's capsule: small, cup-shaped structure in the kidney where filtration occurs

capillary: tiny blood vessel with a wall only one cell thick

cardiac muscle: type of muscle tissue found in the wall of the heart

circulation: the pathway of blood through the body

coronary circulation: the pathway that supplies blood to the heart itself

cortex: the firm outer part of the kidney

dermis: the thick layer of tough fibrous connective tissue that lies under the epidermis of the skin

diastole: the second phase of the complete cycle of heart activity in which the ventricles of the heart relax and receive blood from the atria

epidermis: the outer portion of the skin made up of epithelial cells

erythrocyte: red blood cell that contains hemoglobin

excretion: the process of eliminating nitrogenous wastes from the body

excretory system: the body system that includes the kidneys, bladder, and urethra

fibrin: insoluble protein made up of tiny threads; forms a blood clot

fibrinogen: soluble blood protein that is changed by thrombin into fibrin

glomerulus: the capillary bed that lies inside the Bowman's capsule

heart: large muscular organ that controls the circulation of blood

hemoglobin: pigment found in red blood cells

homeostasis: steady state producing a constant internal environment

inferior vena cava: the large vein that returns blood from the lower part of the body to the heart

kidney: the main excretory organ in humans

leucocyte: white blood cell

lymph: tissue fluid that circulates in the body in special vessels

lymph node: accumulation of lymphatic tissue found at intervals in the course of lymph vessels

lymphatic system: system of thin-walled tubes that conduct lymph from the tissues into the circulatory system

medulla: the inner two-thirds of the kidney

nephron: the unit in the kidney that controls the chemical makeup of blood

pericardium: sac enclosing the heart

perspiration: the excretion of water, salts, and some urea through the skin

plasma: the liquid part of blood

platelet: round disk found in blood, assists in clotting

portal circulation: the pathway of blood vessels that supply the liver

pulmonary artery: artery that carries blood from the heart to the lung

pulmonary circulation: the pathway of blood between the heart and the lungs

pulmonary vein: vein that carries blood from the lung to the heart

pulse: wave of increased pressure passing outwards from the heart along the arteries every time the ventricles contract

receptor: nerve that responds to stimuli from outside or inside the body

red blood cell: blood cell that contains hemoglobin; erythrocyte

renal artery: artery that branches from the aorta and supplies the kidney with blood

renal circulation: the pathway of blood between the heart and the kidneys

renal vein: vein that returns blood from the kidney to the inferior vena cava

Rh factor: type of protein found in red blood cells; its absence in a mother and presence in a fetus may cause a problem in childbearing

semilunar valve: cup-shaped valve located at the opening of an artery

septum: wall separating the two sides of the heart

serum albumin: component of plasma important for regulating osmotic pressure between plasma and blood

serum globulin: component of plasma that contains the antibodies that help fight various diseases

sinoatrial node: small mass of specialized muscle tissue located in the right atrium in which the heartbeat originates

skin: excretory organ of the body made up of the epidermis and the dermis

subcutaneous tissue: layer of fat cells found beneath the skin

superior vena cava: large vein that returns blood to the heart from the upper part of the body

systemic circulation: pathway of blood from the heart to the body cells except to the lungs

systole: first phase of a complete cycle of heart activity in which the ventricles of the heart contract and force blood into the arteries

thrombin: enzyme that changes fibrinogen into fibrin

tonsil: mass of lymphatic tissue found in the throat

ureter: tube that connects the kidney to the bladder

urethra: tube that carries urine from the urinary bladder to the exterior

urine: solution of waste metabolic products produced in the kidneys

vein: blood vessel that carries blood to the heart

ventricle: chamber of the heart that receives blood from the atrium

venule: small vein that is the result of the uniting of capillaries

white blood cell: large nucleated blood cell; leucocyte

Name _____   Date _____

Class _____

HUMAN BODY PLAN—CIRCULATORY AND EXCRETORY SYSTEMS: BASIC TERMS CROSSWORD PUZZLE

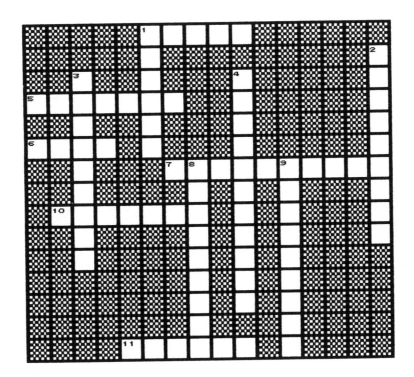

ACROSS

1. Large artery from the heart
5. Urine is collected here
6. Blood vessel carrying blood to the heart
7. Pigment in red blood cells
10. _____ muscle is found in the heart
11. Main excretory organ

DOWN

1. Blood vessel carrying blood away from the heart
2. _____ circulation is between the heart and the lungs
3. Tiny blood vessel with very thin wall
4. Steady state; produces constant internal environment
8. Eliminating wastes
9. White blood cell

HUMAN BODY PLAN—CIRCULATORY AND EXCRETORY SYSTEMS: ADVANCED TERMS CROSSWORD PUZZLE

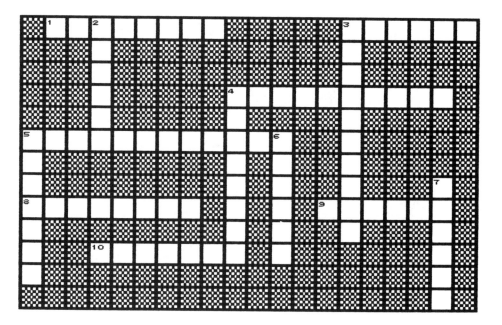

ACROSS

1. Stage in which ventricles relax and receive blood from the atria
3. Tiny insoluble protein threads that form a blood clot
4. Capillary bed in Bowman's capsule
5. _____ tissue is found under the skin
8. Enzyme that changes fibrinogen to fibrin
9. Small vein
10. Serum _____ helps regulate osmotic pressure

DOWN

2. The _____ arch is just above the heart
3. Soluble blood protein that changes to fibrin in order to clot blood
4. Serum _____ contains antibodies
5. Ventricles contract in this phase to force blood into arteries
6. Wall that separates two sides of the heart
7. Layer of fibrous tissue under the epidermis

© 1991 by Center for Applied Research in Education

Name _____ Date _____

Class _____

HUMAN BODY PLAN—CIRCULATORY AND EXCRETORY SYSTEMS: VOCABULARY WORDSEARCH

The wordsearch below contains terms related to our study of the circulatory and excretory systems. The words can be found horizontally in either direction, vertically in either direction, and diagonally in either direction. Clues are given to help you find the words.

```
F K D A Z K O S R J Z Q X L N J M J U L
K K G Y X X Y D P A O R T I C S O F D S
B T L B Q T F A X D G S E P T U M A H H
B Q D O J C R N E P H R O N J D E G U O
T W T P X O Q T J N O F I E H C D C V I
I H L V H M U I D R A C I R E P C T U M
H R R M Y R F W E M L Z R F J G I C W V
N N H O U R B L D F L V T O V T G O A U
S J W T M Y E B V A I F D R B H W P M Y
Y Z P S C B K T N Q C O R O N A R Y B G
E T R L U D I E R E H Y O U A J B R N P
G X U D I P R N Q A P W M J U J C E I N
N N I E V W D Z U Z X S L H E H G S R X
K J J S A U D M V G B R D H N J I K B J
J D X N M Q K P U L M O N A R Y F K I X
G Y G W I X P F T D T Z J B H S C J F K
F Z I E T A N I T U L G G A X O J X J I
R B Z T Y B F X M C E W Z M Z S W F X B
L D E I B A L A V T C Q D D Y D F A N N
E M S B M L G J G D X H W J I Y H X H S
```

CLUES

1. Enzyme that changes fibrinogen into fibrin
2. Insoluble protein that forms blood clot
3. The _____ arch is directly above the heart
4. Blood vessel that carries blood away from the heart
5. Wall separating the two sides of the heart
6. _____ circulation is the system providing blood to the heart itself
7. To clump
8. The _____ vein returns blood from the kidney
9. The sac enclosing the heart
10. The _____ artery brings blood to the lung
11. The blood vessel that carries blood to the heart
12. Unit of the kidney

VOCABULARY

Basic

air pressure
alveolus
basal metabolism
breathing
bronchiole
bronchus
carbon dioxide
carbon monoxide poisoning
diaphragm
epiglottis
expiration
hemoglobin
inspiration
larynx

lung
metabolism
mucous membrane
nasal passage
nostril
oxygen
oxygen transport
pharynx
respiration
thoracic cavity
vital capacity
vocal cord
voice box

Advanced

(all of the basic vocabulary)
anabolism
catabolism
direct respiration
external respiration
indirect respiration
internal respiration
pleural membrane

DEFINITIONS

air pressure: the concentration of air molecules inside or outside the body

alveolus: tiny air sac found in the lung

anabolism: the phase of metabolism that synthesizes carbohydrates, fats, and proteins

basal metabolism: the energy expenditure required to maintain life's basic activities

breathing: the movement of air into and out of the lungs

bronchiole: small branch of the bronchus

bronchus: one of two branches of the lower end of the trachea that leads to the lungs

carbon dioxide: waste product of cellular respiration, CO_2

carbon monoxide poisoning: condition that occurs when carbon monoxide (CO) instead of oxygen binds with hemoglobin

catabolism: the phase of metabolism that involves the oxidation of food molecules and energy release

diaphragm: muscle that separates the abdominal and thoracic cavities and is involved in breathing

direct respiration: the simple exchange of gases through a membrane

epiglottis: flap of cartilage located at the upper end of the trachea

expiration: the discharge of air from the lungs

external respiration: the first stage of indirect respiration when gases are exchanged between the blood and the environment

hemoglobin: the oxygen-carrying pigment found in red blood cells

indirect respiration: the process by which gases enter the bodies and cells of multi-cellular animals

inspiration: taking of air into the lungs

internal respiration: the exchange of gases between the blood and the cells of the body

larynx: voice box

lung: major organ of the respiratory system

metabolism: total of all the chemical processes of a cell or an organism

mucous membrane: form of epithelial tissue that lines body openings and digestive tract and secretes mucus

nasal passage: passage for air that lies above the mouth cavity and connects the nostril to the pharynx

nostril: one of two openings in the nose

oxygen: element that is required for cellular respiration to occur

oxygen transport: process in which oxygen is carried from the lungs to the body cells

pharynx: throat

pleural membrane: membrane that covers the lungs

respiration: process in which glucose is broken down and energy is released

thoracic cavity: body cavity that contains the lungs, trachea, heart, and esophagus

vital capacity: volume of air that can be expelled after full inspiration

vocal cord: structure in the larynx that produces sound

voice box: larynx

HUMAN BODY PLAN—RESPIRATORY SYSTEM:
BASIC TERMS CROSSWORD PUZZLE

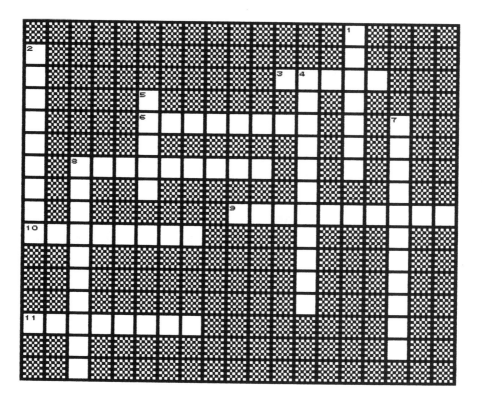

ACROSS

3. Your _____ capacity is the maximum amount of air that you can inhale and exhale
6. Tiny air sac in the lung
8. Movement of air into and out of the lungs
9. Sum of all chemical reactions of a cell or organism
10. Carbon _____ poisoning occurs when carbon _____ binds with hemoglobin
11. A branch of the trachea

DOWN

1. Throat
2. Breathing muscle
4. Taking air into the lungs
5. _____ metabolism indicates the energy needed for life
7. Breakdown of glucose-releasing energy
8. Small branch of the bronchus

© 1991 by Center for Applied Research in Education

Name _____   Date _____

Class _____

HUMAN BODY PLAN—RESPIRATORY SYSTEM: ADVANCED TERMS CROSSWORD PUZZLE

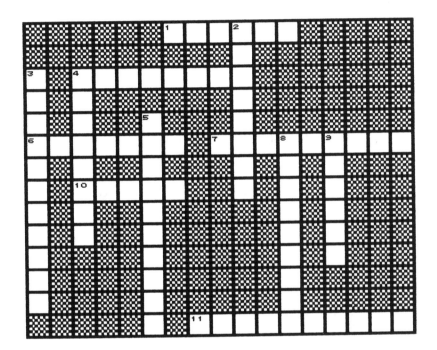

ACROSS

1. _____ respiration is the simple exchange of gases through a membrane
4. _____ respiration is the passage of gases into cells
6. The _____ membrane covers the lung
7. Synthesis of carbohydrates, fats, and proteins
10. The _____ passage connects the nostril to the pharynx
11. Oxygen-carrying pigment

DOWN

2. _____ respiration is gas exchange between the blood and the environment
3. Chemical process of breaking down glucose to release energy
4. _____ respiration is the exchange of gases between the blood and the cells of the body
5. Oxidation of food molecules with energy release
8. Mechanical process of air movement into and out of the lungs
9. Voice box

HUMAN BODY PLAN—RESPIRATORY SYSTEM:
VOCABULARY WORDSEARCH

The wordsearch below contains terms related to our study of the respiratory system. The words can be found horizontally in either direction, vertically in either direction, and diagonally in either direction. Clues are given to help you find the words.

© 1991 by Center for Applied Research in Education

```
P I I F N W E C R Y N Q L F V H T T B O
L P I H D V G O P D Z N T L E C M A R R
F J Q Y B R J S U L O E V L A P E F E Q
F P Q Z Z H M P Z U V H V N S G Q O A T
G I I N D Q V F P I D A O K J Y F T T L
J D R R R T C O B G K I T Y R B G G H R
F I I R T V X M X I T Q A G H K R B I S
H A C V F O M B O A M E O Y B Y G R N K
H P G R F Z V L R Z G M J U E T W O G H
I H V Y L P C I P Z W E U N X H I N P B
F R R R F X P T E Y M T A E C O B C E E
Q A A J N S N H U L S A N O V R J H P K
L G I X E W L U P W I B A C R A Z I Q P
V M Z R N F I C H M L O B U L C X O R W
I A I T K Y A K X G O L O B P I N L K B
H B N M M F R Z X M B I L W Y C T E R W
T Q H G J O Q A D N A S I S Y O K A M Q
Q P L E U R A L L L T M S A Z N H F T W
L Q K B I A T K D W A T M S P D F R N M
G C S E B B K Z J F C L Y C S L F Y P N
```

CLUES

1. The _____ membrane covers the lungs
2. The _____ cavity contains the lungs and the heart
3. Total of an organism's chemical processes
4. Small branch of a bronchus
5. Phase of metabolism that involves oxidation of food and energy release
6. Process of releasing energy from glucose

7. Voice box
8. Tiny air sac in the lung
9. Phase of metabolism that synthesizes organic compounds
10. Mechanical process of moving air into and out of the lungs
11. Breathing muscle that separates the abdominal cavity from the thoracic cavity

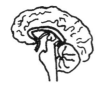 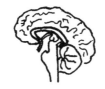

VOCABULARY

Basic

active transport
auditory canal
autonomic nervous system
axon
brain
brainstem
cell body
central nervous system
cerebellum
cerebrum
cochlea
dendrite
eardrum
eustachian tube
forebrain
gray matter
impulse
medulla oblongata
motor nerve
motor neuron
myelin
neuron
neurotransmitter

outer (external) ear
parasympathetic nervous
　system
passive transport
peripheral nervous system
pons
receptor
reflex
response
semicircular canal
sense organ
sensory nerve
sensory neuron
sodium
sodium-potassium pump
somatic nervous system
spinal cord
stimulus
sympathetic nervous system
synapse
taste bud
tympanic membrane
white matter

Advanced

(all of the basic vocabulary)
acetylcholine
action potential
cerebrospinal fluid
convolution
frequency
frontal lobe
ganglion
hemisphere
meninges
parietal lobe
polarized
resting potential
temporal lobe
threshold stimulus

DEFINITIONS

acetylcholine:　chemical substance that transmits the impulse to the muscle fibers to begin the process of contraction

action potential:　wave of depolarization that is the actual nerve impulse that sweeps along the neuron

active transport:　the passage of a substance across a semipermeable membrane requiring the use of energy

auditory canal:　the part of the ear between the outer ear and the eardrum

autonomic nervous system:　division of the nervous system that regulates the internal organs

axon:　fiber that carries impulses away from a nerve cell body

brain:　organ responsible for the integration and control of physiological activities

brainstem: the most posterior portion of the vertebrate brain; includes medulla and midbrain

cell body: part of the nerve cell that contains the nucleus

central nervous system: the brain, spinal cord, and the nerves that come from them

cerebellum: the region of the brain concerned with balance and muscular coordination

cerebrospinal fluid: clear liquid protecting the brain and spinal cord from physical impact

cerebrum: the largest region of the brain; the seat of emotions, intelligence, and voluntary muscular activity

cochlea: spiral passage in the inner ear; its inner surface is lined with nerve endings

convolution: deep wrinkles and furrows in the outer layer of the cerebrum

dendrite: nerve fiber that carries impulses toward a nerve cell body

eardrum: membrane in the ear; caused to vibrate by sound waves

eustachian tube: canal that connects the mouth cavity with the middle ear

forebrain: cerebrum

frequency: the number of impulses passing through a nerve fiber each second

frontal lobe: one of two sections of the cerebrum responsible for personality, emotion, judgment, and self-control

ganglion: mass of neuron cell bodies located outside the central nervous system

gray matter: tissue of the cerebrum made up of large numbers of neuron cell bodies

hemisphere: one of two halves of the cerebrum

impulse: electrochemical charge moving along a neuron

medulla oblongata: the region of the brain that controls the activities of the internal organs

meninges: three membranes covering the brain and spinal cord

motor nerve: nerve that carries impulses from the brain or spinal cord to a muscle or a gland

motor neuron: motor nerve

myelin: fatty insulating substance that surrounds the axons of many motor neurons

neuron: nerve cell

neurotransmitter: chemical stored in small sacs at the end of axons

outer (external) ear: the part of the ear located outside of the head

parasympathetic nervous system: the part of the nervous system that dominates when conditions are not stressful

parietal lobe: one of two lobes of the cerebrum that interprets sensations of pain, pressure, touch, heat and cold, and position in space

passive transport: movement of molecules across a membrane without energy being expended

peripheral nervous system: the nervous system that communicates with the central nervous system by a system of nerves

polarized: condition in which ions of opposite charges are separated by a semipermeable membrane in nerve fibers

pons: the upper portion of the brain stem that receives stimuli from the facial area

receptor: specialized sense organ that receives stimuli

reflex: autonomic response to a stimulus

response: reaction to a stimulus

resting potential: the difference in charge across a nerve cell membrane

semicircular canal: one of three loop-shaped tubes in the inner ear that functions in balance

sense organ: organ that contains specialized receptors

sensory nerve: nerve cell that carries sensory information toward the central nervous system

sensory neuron: sensory nerve

sodium: positive ion located in excess amounts on the outside of the nerve cell

sodium-potassium pump: an active transport mechanism that keeps the sodium ion concentration high outside the cell

somatic nervous system: axons that cause contractions of muscle cells

spinal cord: the main nerve of the central nervous system that extends down the back

stimulus: factor or environmental change capable of producing activity

sympathetic nervous system: division of the autonomic nervous system that includes two rows of nerve cords along the spinal column

synapse: the space between two connecting neurons

taste bud: tiny bump located on the surface of the tongue with nerve endings at its base

temporal lobe: one of two lobes of the cerebrum in which the sense of hearing is interpreted

threshold stimulus: the strength of a stimulus sufficient to produce a change in polarity

tympanic membrane: eardrum

white matter: tissue below the cortex of the cerebrum; formed of masses of nerve fibers covered by fatty sheaths

HUMAN BODY PLAN—NERVOUS SYSTEM AND SENSE ORGANS: BASIC TERMS CROSSWORD PUZZLE

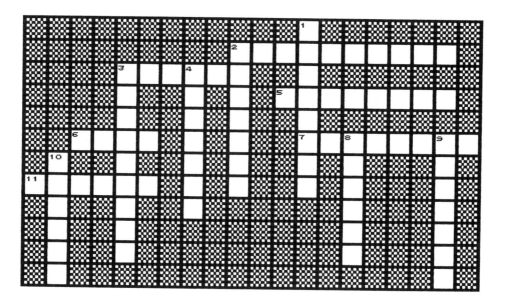

ACROSS

2. Region of brain concerned with balance and muscular coordination
3. _____ transport requires energy
5. Carries impulses toward nerve cell body
6. Carries impulses away from nerve cell body
7. Reaction to stimulus
11. Nerve cell

DOWN

1. Region of brain that is the seat of emotions and intelligence
2. The _____ nervous system is made up of the brain, the spinal cord, and the nerves from each of them
3. The _____ nervous system regulates the internal organs
4. Charge moving along a neuron
8. The _____-potassium pump keeps the sodium concentration high outside of the cell
9. Space between two neurons
10. Autonomic response to a stimulus

HUMAN BODY PLAN—NERVOUS SYSTEM AND SENSE ORGANS: ADVANCED TERMS CROSSWORD PUZZLE

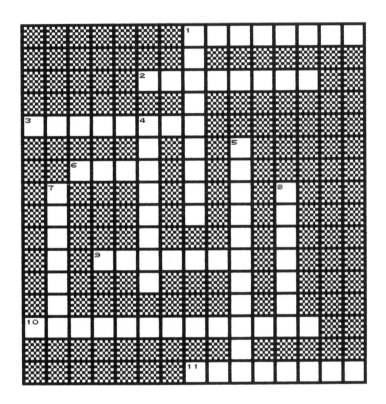

ACROSS

1. Hearing is interpreted in the _____ lobe
2. Pain, pressure, touch, position in space, and temperature are interpreted in the _____ lobe
3. Layers that cover the brain and spinal cord
6. Carries impulses away from the nerve cell body
9. Personality, judgment, emotion, and self-control are governed by the _____ lobe
10. _____ fluid is found in the brain
11. Carries impulses towards the nerve cell body

DOWN

1. A _____ stimulus is just strong enough to be transmitted
4. Neurons located outside of the central nervous system
5. Wrinkle or furrow in the cerebrum
7. Space between neurons
8. Part of the inner ear

HUMAN BODY PLAN—NERVOUS SYSTEM AND SENSE ORGANS: VOCABULARY WORDSEARCH

The wordsearch below contains terms related to our study of the nervous system and sense organs. The words can be found horizontally in either direction, vertically in either direction, and diagonally in either direction. Clues are given to help you find the words.

```
Z Y G F S P A R I E T A L A C T M D C J
K V K D Q C O D O I I R E C P Y Q E C N
W L B L R K E M C Q N S L E A O V N C N
N E W P E Y M P T X P I T T B R D D T F
O G S Q F R C S D A Z Y W Y J J K R N N
I N L N L X M M N H M E N L R U I I I N
L M V G E L X Y A N X R V C E Q L T M N
G F E O X W S M G Z N A I H D Z N E U U
N W F L A R O P M E T X S O W X F I U D
A F N B G Y N T L H C G E L B R J J J Q
G L B R T B O A E N N X H I O A M S H R
S V Q L I H X J T V M W N N U L W E O V
E T Y I R O A Q E C J E T E O C G S O R
M F G C Z C F V X D I A K V X I N L X I
I X H L A O X I T Q L F W M R A F P S K
N D G Q F Q R G O F P Q Q Q H N T X Y B
B Q W H W G X X R W S N Q B X K I Z M F
S T I M U L U S U T H R E S H O L D Q K
Y T E I C E R E B R O S P I N A L J F Q
S P T W T I Z V A A N C D B D T W M S N
```

CLUES

1. Hearing is interpreted in the _____ lobe of the brain
2. Nerve fiber carrying impulses away from nerve cell body
3. Causes response
4. _____ fluid bathes the brain
5. Neurons located outside the central nervous system
6. A nerve's _____ level must be reached before it will send an impulse
7. Part of a nerve carrying impulses towards the nerve cell body
8. Response to a stimulus; autonomic
9. Chemical that transmits impulse to muscle; causes contraction
10. The _____ lobe of the brain is responsible for such things as emotion, judgment, and self-control
11. Sensations such as pain, pressure, and temperature are interpreted in the _____ lobe
12. Space between connecting neurons

VOCABULARY

Basic   *Advanced*

adrenal gland
adrenaline
cortisone
diabetes
duct
endocrine gland
estrogen
exocrine gland
follicle stimulating
    hormone (FSH)
gland
glucagon
gonad
homeostasis
hormone
insulin
luteinizing hormone (LH)

luteotrophic hormone (LTH)
negative feedback
noradrenaline
ovary
pancreas
parathyroid gland
parathyroid hormone
pituitary gland
progesterone
protein
puberty
testicle
testosterone
thymus
thyroid gland
thyroxine

(all of the basic vocabulary)
anterior pituitary
diabetes mellitus
hyperthyroidism
hypothyroidism
oxytocin
posterior pituitary
steroid hormone
vasopressin

DEFINITIONS

adrenal gland: endocrine gland located at the top of the kidney

adrenaline: hormone that causes the "fight or flight" response and is secreted by the medulla of the adrenal gland

anterior pituitary: the front part of the pituitary gland that synthesizes proteins that regulate the secretions of other endocrine glands

cortisone: hormone made by the cortex of the adrenal gland; it functions in cell metabolism, diminishes local inflammation, and helps heal wounds

diabetes: disease caused by insufficient insulin production

diabetes mellitus: disease caused by insufficient insulin production

duct: tube that carries fluids from one system to another

endocrine gland: ductless gland

estrogen: one of a group of female hormones secreted by the ovaries

exocrine gland: gland with ducts

follicle stimulating hormone (FSH): hormone of the anterior pituitary gland that stimulates the growth and maturation of the ovum

gland: organ that makes specific chemical substances (hormones) for secretion

glucagon: polypeptide hormone produced in the pancreas that causes the blood sugar level to rise

gonad: sexual organ or reproductive gland that produces gametes

homeostasis: steady state producing a constant internal environment

hormone: chemical secreted by a gland in one part of the body that controls chemical reaction in another part

hyperthyroidism: condition caused by an overactive thyroid gland

hypothyroidism: condition caused by an underactive thyroid gland

insulin: substance, secreted by the pancreas, that regulates the uptake of glucose from the blood into the cells

luteinizing hormone (LH): pituitary hormone that acts with FSH to cause ovulation and stimulate formation of the corpus luteum

luteotrophic hormone (LTH): hormone secreted by the pituitary that causes the development of breasts, milk production, and the secretion of progesterone by the corpus leteum

negative feedback: mechanism of the endocrine system by which the change in concentration of one hormone balances the production of another hormone

noradrenaline: hormone that causes the "fight or flight" response and is secreted by the medulla of the adrenal gland

ovary: organ in which female gametophytes are produced

oxytocin: pituitary hormone that helps to regulate blood pressure and to stimulate smooth muscles

pancreas: organ that has both digestive and endocrine functions

parathyroid gland: one of four endocrine glands that play a role in bone growth, muscle tone, and nerve activity

parathyroid hormone: hormone produced by the parathyroid gland

pituitary gland: small endocrine gland located at the base of the brain

posterior pituitary: the part of the pituitary gland that secretes oxytocin and vasopressin

progesterone: female hormone secreted by the ovaries

protein: organic compound made up of amino acids

puberty: the stage of life that begins with the maturation of the ovaries in females and testes in males

steroid hormone: complex hormone secreted by the cortex of the adrenal glands

testicle: organ that produces male gametes

testosterone: male hormone produced in the testes

thymus: gland that functions in switching on the immune system at birth

thyroid gland: large endocrine gland in the neck that regulates metabolism

thyroxine: iodine-containing amino acid hormone

vasopressin: hormone secreted by the hypothalamus that acts in maintaining water balance

Name _____   Date _____

                                     Class _____

HUMAN BODY PLAN—ENDOCRINE SYSTEM:
BASIC TERMS CROSSWORD PUZZLE

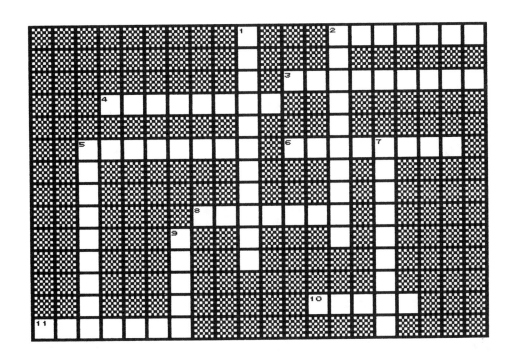

ACROSS

2. The _____ gland is at the top of the kidney
3. Hormone made by the adrenal gland; reduces inflammation
4. Female hormone secreted by ovary
5. Organ with both digestive and endocrine functions
6. Disease caused by insufficient insulin
8. Regulates uptake of glucose
10. Organ that makes substances for secretion
11. The _____ gland is in the neck and regulates metabolism

DOWN

1. Steady state
2. Secreted by adrenal gland; "fight or flight" response
5. Small gland located at the base of the brain
7. _____ glands are ductless
9. Sexual organ

HUMAN BODY PLAN—ENDOCRINE SYSTEM: ADVANCED TERMS CROSSWORD PUZZLE

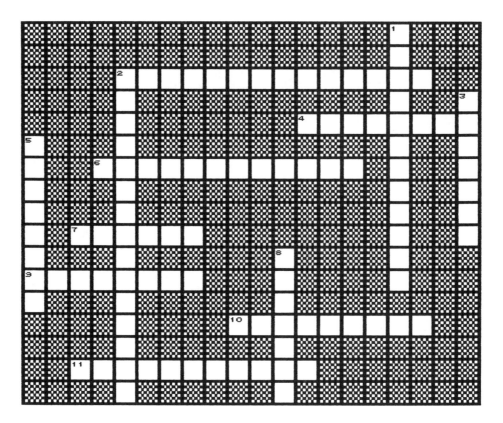

ACROSS

2. Condition caused by underactive thyroid gland
4. Female hormone secreted by ovary
6. Female hormone
7. Gland that switches on immune system
9. Pituitary hormone; helps regulate blood pressure and stimulate smooth muscles
10. Hormone that reduces local inflammation and helps heal
11. Hormone that helps maintain water balance

DOWN

1. Male hormone produced in testes
2. Condition caused by underactive thyroid gland
3. Regulates uptake of glucose
5. The _____ pituitary gland makes proteins that regulate the secretions of other glands
8. _____ hormones are secreted by the cortex of the adrenal glands

Name _____ Date _____

Class _____

HUMAN BODY PLAN—ENDOCRINE SYSTEM: VOCABULARY WORDSEARCH

The wordsearch below contains terms related to our study of the endocrine system. The words can be found horizontally in either direction, vertically in either direction, and diagonally in either direction. Clues are given to help you find the words.

```
E N L B X Y O V A A M Q H N L J B A J C
D U V I N M C D G P G T A Y Q R L S H P
Q K M R J X Z Z E E N I R C O X E Y R N
P Z P G K Y V C C Y E H D B Q E J O O I
I S E T E B A I D Y T B F G A Q G H P T
T X G E N D O C R I N E Z J E E Y W P K
U S V F E I O G C G B J E N S P T I O X
I D V A M N X S R A E I O T O X V C V M
T F B P S T E H K U D S E T Q H Z E Z M
A A S J L O O L W T I R H X J S Y P D N
R O F M T E P G D T O Y U U G I J E C Y
Y U H N M C X R R N R U Q U F V L N Y U
I Z N I U E D O E O E N B P A Z U I L W
F C Z L L Q C D I S T F D E H X J L A Y
K Z W U A N V D R Y S U M W O I F A V U
Q Q Y S Z X I R W J W I Z B L P D N G F
E V Y N R S U I F U B O N M Z G Z E K O
P T V I M B W H X O X Y T O C I N R P W
E P P U H Q Z N F Y R G F K F Z D D J W
A M D B N D F S N E U C H U D R S A J R
```

CLUES

1. Hormone that helps maintain water balance
2. Hormone that helps regulate blood pressure and stimulate smooth muscles
3. Female hormone secreted by the ovaries
4. _____ glands have ducts
5. _____ hormones are secreted by the cortex of the adrenal gland
6. _____ is caused by an underactive thyroid gland
7. Regulates glucose uptake
8. _____ glands have no ducts
9. The _____ gland is a small endocrine gland located at the base of the brain
10. Hormone that causes the "fight or flight" response
11. Disease caused by insufficient insulin
12. Hormone that diminishes local inflammation and helps heal wounds

VOCABULARY

Basic

afterbirth
amnion
clitoris
corpus luteum
Cowper's gland
embryo
epididymis
estrogen
Fallopian tube
fertilization
fetus
follicle stimulating
  hormone (FSH)
implantation
labium
luteinizing hormone (LH)
menstruation
ovary
ovulation
penis

placenta
progesterone
prostate gland
pubic bone
scrotum
seminal vesicle
seminiferous tubule
sexual reproduction
sperm
testicle
umbilical cord
urethra
uterus
vagina
vas deferens
yolk sac
zygote

Advanced

(all of the basic
  vocabulary)
vulva

DEFINITIONS

afterbirth: the placenta and the remains of the amnion that are expelled after a child is born

amnion: the innermost fetal membrane that forms the sac that encloses the fetus

clitoris: small erectile body at the anterior part of the vulva; homologous to the penis

corpus luteum: the ruptured follicle after the ovum has been discharged

Cowper's gland: small male sex organ that releases a secretion that will become part of the semen

embryo: structure formed from the zygote that develops into a fetus

epididymis: male organ where sperm mature and are stored until needed

estrogen: one of a group of female hormones secreted by the ovaries

Fallopian tube: the tube serving to transport the egg to the outside of the uterus

fertilization: the union of a sperm cell and an egg cell

fetus: developmental stage that begins six to eight weeks after fertilization and lasts until birth

follicle stimulating hormone (FSH): hormone of the pituitary that stimulates the growth and maturation of the ovum

implantation: the attachment of the embryo to the lining of the uterus

labium: fold of skin of the vulva, encloses and protects underlying, more delicate structures; homologous to the scrotum

luteinizing hormone (LH): pituitary hormone that acts with FSH to cause ovulation and the secretion of estrogen

menstruation: the breakdown and discharge of the uterine tissue and unfertilized ovarian egg

ovary: organ in which female gametophytes are produced

ovulation: the release of a mature ovum

ovum: the egg cells; the female gamete

penis: the organ by which sperm are introduced into the female

placenta: large spongy tissue in the uterus that transports substances between the mother and the developing young

progesterone: female hormone secreted by the ovaries

prostate gland: gland below the urinary bladder that secretes some of the substances that make up semen

pubic bone: bone in each pelvic girdle

scrotum: pouch of skin that holds the testicles

seminal vesicle: structure that stores sperm cells

seminiferous tubule: tightly coiled tube within the testicle

sexual reproduction: reproduction involving the fusion of two special cells, each having a haploid chromosome number

sperm: male reproductive cell

testicle: male reproductive organ

umbilical cord: structure that connects the developing embryo to the placenta

urethra: tube leading from the urinary bladder to an external opening of the body

uterus: the organ in which young mammals develop before birth

vagina: female organ that receives sperm from the male penis; forms part of the birth canal

vas deferens: ciliated duct through which the sperm pass after leaving the epididymis

vulva: external genitalia in females; in humans includes the clitoris and labia

yolk sac: embryonic membrane that provides food for the embryo

zygote: fertilized egg

HUMAN BODY PLAN—REPRODUCTIVE SYSTEM:
BASIC TERMS CROSSWORD PUZZLE

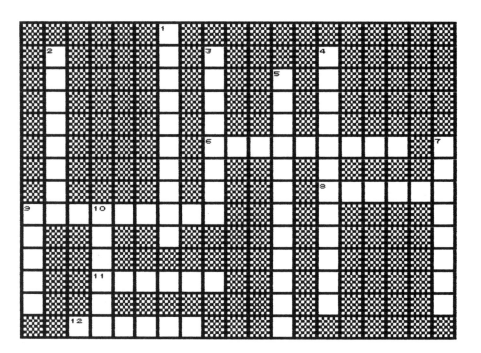

ACROSS

6. Release of a mature ovum
8. Organ in which the fetus develops
9. _____ tubes are also called oviducts
11. The zygote develops into an _____
12. Innermost fetal sac

DOWN

1. Stores sperm
2. Membrane that transports material between mother and fetus
3. Female hormone
4. Discharge of unfertilized egg and uterine tissue
5. Attachment of embryo to uterus
7. Male reproductive organ
9. Developing child from eight weeks until birth
10. The ovum is discharged from the corpus _____

Name _____ Date _____

Class _____

HUMAN BODY PLAN—REPRODUCTIVE SYSTEM: ADVANCED TERMS CROSSWORD PUZZLE

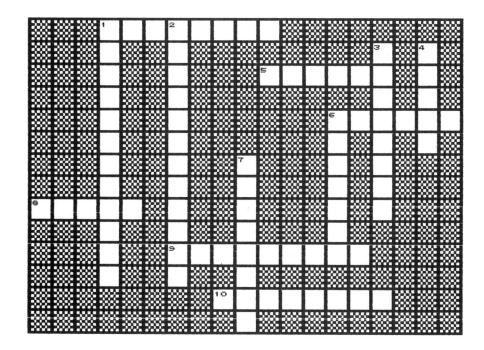

ACROSS

1. Secretes fluids for semen
5. Fertilized egg
6. Organ in which fetus develops
8. Female reproductive organ that produces eggs
9. The _____ cord connects the embryo with the placenta
10. Spongy structure that transports materials between fetus and mother

DOWN

1. Female hormone
2. The _____ tubule is in the testicle
3. The sperm pass through the vas _____
4. Developing baby
6. Tube from urinary bladder to outside the body
7. Male reproductive organ

HUMAN BODY PLAN—REPRODUCTIVE SYSTEM:
VOCABULARY WORDSEARCH

The wordsearch below contains terms related to our study of the reproductive system. The words can be found horizontally in either direction, vertically in either direction, and diagonally in either direction. Clues are given to help you find the words.

```
A Y C P H Z Z T P S D H O C Y V B W C A
N W A W J N J L E N N Y Z N H T C F D X
Z C V I Q K A V S I R O A D A N I Z V I
H K N M V C A E W P M L A I H N N F E Q
Q O X D E M M H D R C Z C T P Q A Q W H
H V U N C N N D G O U S G A W Q D I X H
F E T Y Q E I Z I G L D V T L E K P R G
S A A M B G O U L E H A P N N L U Q L O
U G F C E O N M Q S G D Z A V R M N I M
N N W X C R M R Z T N W I L G M O S J J
R T U X D T M M L E F M E P B E Z Z K Q
E E Y J D S H T X R D R R M G E U S H Q
G Y T M Z E H P J O P E R I I T B G R Z
D F T O T E A J T N L P Y F E W E N A P
L J W R G S W T D E N S L R S U T E F O
O K C Y S Y M C I Q U V U X W Y Z M X K
E Q Q D G H Z L Y O Q S H C Z B B P V I
C H H L L U T E I N I Z I N G P M R I U
S E O V U L A T I O N B B Y A R G X M B
L C N F E R T I L I Z A T I O N Q Z F A
```

CLUES

1. Fertilized egg
2. Female hormone secreted by ovaries
3. Attachment of embryo to uterus
4. Innermost fetal membrane; forms a sac that encloses the fetus
5. Organ in which young mammals develop
6. Structure that transports materials between mother and fetus
7. Unborn mammal; in humans, the stage from about six weeks after conception until birth

8. _____ hormone, abbreviated LH, acts with FSH to cause ovulation
9. Male gamete
10. Release of mature ovum
11. Female hormone
12. Union of sperm and egg

VOCABULARY

Basic

absorption
addiction
alcohol
amphetamine
analgesic
barbiturate
caffeine
cancer
carbon monoxide
cirrhosis
cocaine
codeine
crack
depressant
downer
drug
emphysema
gastritus
hallucinogen

heroin
inhalent
LSD
marijuana
morphine
narcotic
nicotine
opium
over-the-counter drug
PCP
psychoactive drug
reflex
sedative
stimulant
tar
tranquilizer
upper
Valium™

DEFINITIONS

absorption: the process in which chemicals are taken in by the body cells

addiction: drug dependence; caused by use over a period of time

alcohol: depressant produced by the action of yeast on the sugars in fruits or grains

amphetamine: central nervous system stimulant; "pep pill"

analgesic: a drug that relieves pain

barbiturate: synthetic drug that acts as a tranquilizer or sedative

caffeine: natural drug found in coffee, tea, cocoa, and the cola nut

cancer: any malignant growth or tumor

carbon monoxide: poisonous gas found in cigarette smoke; CO

cirrhosis: liver disease that can be caused by excessive use of alcohol over an extended period of time

cocaine: central nervous system stimulant derived from the leaves of the coca plant

codeine: narcotic often used in cough medicines

crack: concentrated form of cocaine

depressant: drug that decreases the activity of the central nervous system

downer: depressant

drug: any substance taken into the body that alters its normal processes

emphysema: degenerative lung disease that interferes with the exchange of gases in the lungs

gastritus: painful swelling of the stomach lining

hallucinogen: drug that changes sensory perception, affects perceptions of time and space, and influences the content of the user's thoughts

heroin: narcotic derived from morphine

inhalent: volatile substance that causes dizziness and loss of coordination when breathed in

LSD: hallucinogenic drug that may elevate heart rate and blood pressure; lysergic acid diethylamide

marijuana: an intoxicant made of the flowers, leaves, and seeds of the Indian hemp plant

morphine: a narcotic depressant and painkiller derives from opium

narcotic: habit-forming central nervous system depressant

nicotine: toxic, addicting substance found in tobacco

opium: substance obtained from the juice of the white poppy, used to manufacture narcotics

over-the-counter drug: any drug that can be obtained without a doctor's prescription

PCP: drug that has been known to act as a stimulant, a hallucinogen, and an analgesic; "angel dust"

psychoactive drug: any drug that affects the central nervous system

reflex: automatic response to a stimulus

sedative: drug with a soothing or tranquilizing effect

stimulant: drug that increases the activity of a body sytem

tar: residue of cigarette smoke; a thick substance that can settle on the mucous membranes and cause irritation

tranquilizer: drug that acts to reduce anxiety and depression

upper: stimulant

Valium™: frequently prescribed tranquilizer; diazepam

Name _____   Date _____

Class _____

TOBACCO, ALCOHOL, AND OTHER DRUGS:
BASIC TERMS CROSSWORD PUZZLE 1

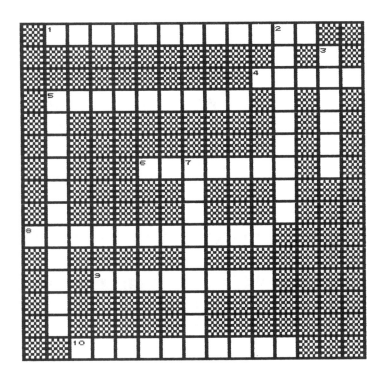

ACROSS

1. Changes sensory perception
4. Stimulant
5. Dependence
6. Drug made from coca plant's leaves
8. Synthetic sedative
9. Addictive depressant
10. Decreases activity of central nervous system

DOWN

2. Lung disease
3. Automatic response to stimulus
5. Pep pill
7. Liver disease caused by alcohol

TOBACCO, ALCOHOL, AND OTHER DRUGS:
BASIC TERMS CROSSWORD PUZZLE 2

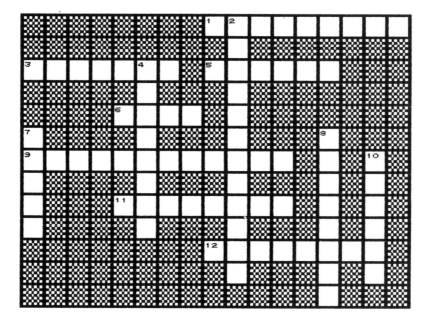

© 1991 by Center for Applied Research in Education

ACROSS

1. Increases activity of central nervous system
3. A narcotic depressant present in some cough medicines
5. A tranquilizer often abused
6. A substance taken into the body that alters normal body functions
9. Any drug that affects the central nervous system
11. Toxin in tobacco
12. Drug causing sleepiness

DOWN

2. Drug used to treat anxiety and depression
4. Physically addictive depressant
7. Substance from a poppy
8. Painkiller derived from opium
10. Depressant

Name _____ Date _____

Class _____

TOBACCO, ALCOHOL, AND OTHER DRUGS:
VOCABULARY WORDSEARCH

The wordsearch below contains terms related to our study of tobacco, alcohol, and drugs. The words can be found horizontally in either direction, vertically in either direction, and diagonally in either direction. Clues are given to help you find the words.

```
F I S V G L A A S Q M I W C Y I T F K N
Z W J D J X Q I O X Y U Q V L T A M M N
S S I H O I W N I R R L C Z H L N Z B T
K D Y Z D L Z H M S E D X X D E Z M T P
B J S H F D W Y Q R F T H W G B P U U Q
W G I L U C J H M M N Z B O E L J Q Y A
A K S J N A R T C A H P N N I V T M A Y
Y K O R G N Y W S W G I I Q E U O D M Y
Z J H Q O P S S G A C T O V W E D N P F
Y C R X D Q E N S U O Z H I V I P A H H
I A R U X R X T L C W H H I C Z T R E X
G N I C P O R L I T L Q T T S P V C T X
R G C E P I A N Y B F C I E Q Z Z O A B
A O D F T H S Q O G A O D U C G V T M U
E M L I T O E T F O N A J T A S P I I X
F B S F C U G Y H G T W S J N N Z C N O
H J B B B H H C W I H O J N C P G C E T
I F M O Y J Y E V P C J A H E Y V I A T
D E V O F S Q E U G Z T P N R D C U I T
D R V B P B R A M Q G P Q W I X S J P Z
```

CLUES

1. Drug that causes sleepiness
2. Potentially physically addictive drug
3. Painful swelling of stomach lining
4. Malignant growth or tumor
5. Drug that affects the central nervous system
6. Hallucinogenic drug that may elevate heart rate and blood pressure
7. Drug that decreases central nervous system activity

8. Stimulant
9. Toxin found in tobacco
10. Chemical that changes sensory perception, affects perceptions of time and space, and influences thoughts
11. Liver disease
12. Dependence on drug caused by use over a period of time

VOCABULARY

Basic

		Advanced
abiotic factor	grassland	(all of the basic vocabulary)
biome	interaction	benthic zone
biosphere	intertidal zone	denitrification
biotic factor	limiting factor	littoral zone
canopy	marine biome	nitrification
carbon-oxygen cycle	nitrogen cycle	pelagic zone
climax vegetation	ocean	xerophyte
coniferous forest	plankton	
deciduous forest	precipitation	
desert	predation	
ecology	predator	
environment	prey	
epiphyte	rain forest	
estuary	resource	
evaporation	terrestrial biome	
food chain	tundra	
freshwater biome	upwelling	
	water cycle	

DEFINITIONS

abiotic factor: any element of the environment that is nonliving

benthic zone: the bottom, or floor, of an aquatic environment

biome: large natural area that has a particular climate or physical condition

biosphere: the part of the earth and its atmosphere inhabited by living organisms

biotic factor: any living organism in an environment

canopy: the highest level of vegetation in a terrestrial biome

carbon-oxygen cycle: cycle that includes the two basic life processes of respiration and photosynthesis

climax vegetation: group of plants that dominates a terrestrial biome

coniferous forest: biome in which the climax vegetation consists of cone-bearing trees

deciduous forest: temperate biome with a long growing season and in which the climax vegetation consists of deciduous trees

denitrification: the removal of nitrogen by bacteria from the nitrogen cycle

desert: extremely dry biome in which the climax vegetation consists of plants that have well-developed root systems

ecology: the study of the interaction of living organisms with their environment

environment: all of the living and nonliving things that act upon an organism

epiphyte: organism that grows on another organism without hurting it

estuary: zone between freshwater and marine biomes

evaporation: of water, change of phase from liquid to gas

food chain: series of organisms in any community in which each member feeds on the preceding member and is in turn eaten by a subsequent member

freshwater biome: any biome including lakes, ponds, inland swamps, springs, streams, and rivers, characterized by presence of fresh, not salt water

grassland: biome in which the climax vegetation consists of many species of grasses

interaction: any relationship between living and nonliving things

intertidal zone: shore area that is alternately covered and uncovered by water

limiting factor: resource for which demand is greater than supply

littoral zone: the area of a lake from the water's edge to a point where light penetrates

marine biome: biome that includes the oceans and seas, characterized by presence of salt water

nitrification: the conversion of organic nitrogen compounds into nitrates that are used by plants

nitrogen cycle: the circulation of nitrogen in nature

ocean: large biome of salt water

pelagic zone: that zone of a marine biome above the benthic zone

plankton: basic aquatic food source including diatoms, dinoflagellates, unicellular algae, protozoans, and the larval forms of many animals

precipitation: rain, snow, hail, sleet, or mist, or settling out of a solid from a mixture in liquid

predation: the act of one animal feeding upon another

predator: animal that feeds upon another animal

prey: animal killed by a predator

rain forest: biome with abundant water supply, long growing season, warm temperature, and plentiful plant and animal life

resource: any substance available from the environment

terrestrial biome: land biome

tundra: biome of arctic and subarctic regions characterized by a very cold and dry climate

upwelling: phenomenon that occurs when surface currents and differences in water temperatures cause deeper waters to be brought to the surface

water cycle: the circulation of water molecules in nature

xerophyte: plant that lives in dry habitats such as deserts and that can withstand prolonged drought

Name _____ Date _____

Class _____

INTRODUCTION TO ECOLOGY:
BASIC TERMS CROSSWORD PUZZLE

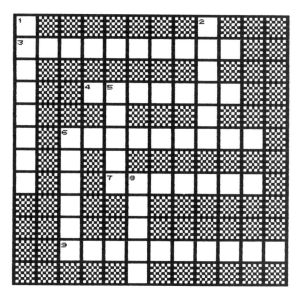

ACROSS

3. The _____ zone is exposed at low tide
4. _____ factors include air and water
6. Part of earth and atmosphere in which life exists
7. Study of organisms and their interaction with their environments
9. The _____ forest consists mostly of gymnosperms

DOWN

1. The _____ factor is whatever an organism has too much of or too little of to live in a place
2. Biome, includes oceans
5. Major type of environment
6. _____ factors are organisms
8. A food _____ is a sequence of organisms that eat and are in turn eaten by others

Name _____ Date _____

Class _____

INTRODUCTION TO ECOLOGY:
ADVANCED TERMS CROSSWORD PUZZLE

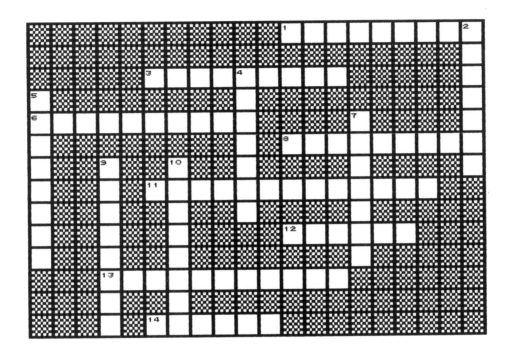

ACROSS

1. Part of earth and its atmosphere where organisms live
3. Desert plant
6. Zone between high and low tides
8. Brings deep water to surface
11. Forms nitrates for use by plants
12. Biome, includes oceans
13. Land biome
14. The _____ vegetation is the kind of plant that dominates a biome

DOWN

2. Study of organisms and their environment
4. Open ocean zone
5. Nitrogen moves through the biosphere in the _____ cycle
7. Zone on bottom of the ocean
9. Floating organisms
10. Shallow water zone

INTRODUCTION TO ECOLOGY:
VOCABULARY WORDSEARCH

The wordsearch below contains terms related to our study of ecology. The words can be found horizontally in either direction, vertically in either direction, and diagonally in either direction. Clues are given to help you find the words.

```
D G J E E G N C A M G O L L R C H M F S
M C X A M I L C Q M H E I A A N G W A U
I T F S D L L E Q Q J S T N P G K C C Q
A L R D P P I K I H N U T M T Y U I Y Y
S H Q H Y G O L O C E P O F H Y E T A D
K O A W E R V F A J E E R N A S Z O U A
C S V D W K L E E N U O A E R R O I S J
R C O N I F E R O U S P L P K L Z B P J
O E V R G L N U D H F S L E F T C A U O
T G Y L T L V V T S D A A U E M T D A Q
J M Z Q Z I N A G V N R X T D B G K Y O
C E T J S N A Y X K U H Y S M I N Y N G
O S R S P C D X T M X H C U Q J I R F Z
X G L E Y R J O H H P J I O U K T A K Z
G T P B H J N T Q O M F G U F A I U F U
X W T F F P A B R Q T F A D U C M T P V
O S M A O G S E Q P L R L I Q M I S S E
A Y O C H U X O I Z E R E C B M L E V E
I X P H T W E R I L P T P E K T J V C T
E N T L D V L H H B W I D D K W T V J E
```

© 1991 by Center for Applied Research in Education

CLUES

1. Plant that lives in dry areas
2. Small floating organism
3. Study of interactions of organisms and their environment
4. _____ vegetation dominates a terrestrial biome
5. Marine zone above benthic zone; open water
6. _____ factors are the resources that an organism runs out of first
7. A _____ forest consists mostly of trees that lose their leaves in the winter
8. Area of earth inhabited by organisms
9. The _____ zone is found along the shore
10. Zone between fresh water and marine biomes
11. A _____ forest is dominated by cone-bearing trees
12. An _____ factor is nonliving

1: ORGANIC CHEMISTRY: ANSWER KEY

BASIC TERMS CROSSWORD PUZZLE

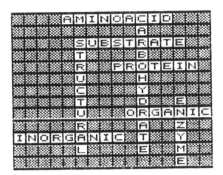

ADVANCED TERMS CROSSWORD PUZZLE

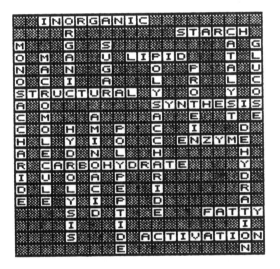

VOCABULARY WORDSEARCH

2: CELL STRUCTURE AND FUNCTION: ANSWER KEY

BASIC TERMS CROSSWORD PUZZLE

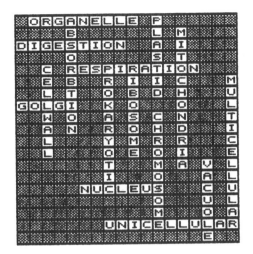

ADVANCED TERMS CROSSWORD PUZZLE

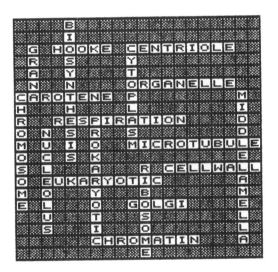

VOCABULARY WORDSEARCH

3: ENERGY, RESPIRATION, AND PHOTOSYNTHESIS: ANSWER KEY

BASIC TERMS CROSSWORD PUZZLE

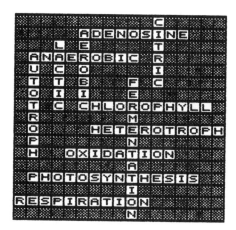

ADVANCED TERMS CROSSWORD PUZZLE

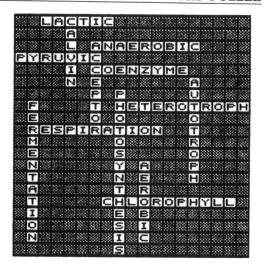

VOCABULARY WORDSEARCH

4: NUCLEIC ACIDS AND PROTEIN SYNTHESIS: ANSWER KEY

BASIC TERMS CROSSWORD PUZZLE

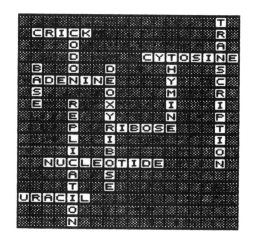

ADVANCED TERMS CROSSWORD PUZZLE

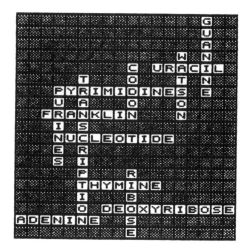

VOCABULARY WORDSEARCH

5: CELL REPRODUCTION AND GROWTH: ANSWER KEY

BASIC TERMS CROSSWORD PUZZLE

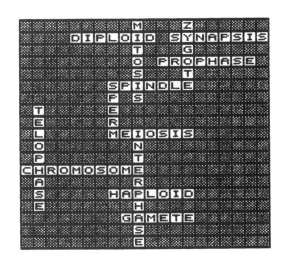

ADVANCED TERMS CROSSWORD PUZZLE

VOCABULARY WORDSEARCH

6: PRINCIPLES OF HEREDITY: ANSWER KEY

BASIC TERMS CROSSWORD PUZZLE

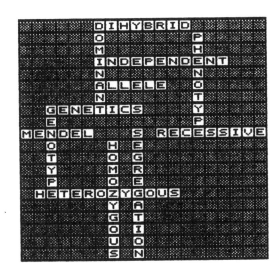

ADVANCED TERMS CROSSWORD PUZZLE

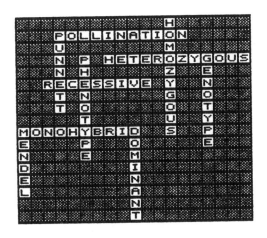

VOCABULARY WORDSEARCH

7: MECHANISMS OF HEREDITY: ANSWER KEY

BASIC TERMS CROSSWORD PUZZLE

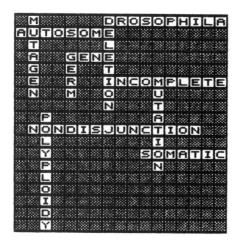

ADVANCED TERMS CROSSWORD PUZZLE

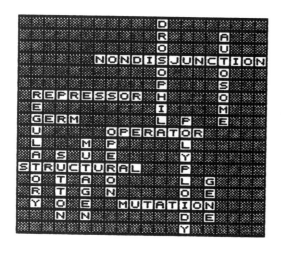

VOCABULARY WORDSEARCH

```
. . . . . . . . G N I P P A M . . . . . . .
. . . . . . . . . T N A T U M . . . S . .
. . . . . . N E G A T U M . . . . E . .
N P O L Y P L O I D Y . . . . M . . .
O . E . R . . D E K N I L X E S O . . . . .
N . E . E . . . . . . . . . . S . . . . .
D . I . G . . . . . . . . . O . . . . .
I . E . U . . . . M . . . . . . . . .
S . G . L . . D R O S O P H I L A . .
J . E . A . . . R M U T A T I O N . E
U . N . T . . . H . . . . . . . . T .
N . M . O . . C R . . . . . . . . E .
C . O . R . X E M O S O M O R H C Y L .
T . S . Y . E . . . S . . . . . . P .
I . O . . S . . . . S . . . . . . M .
O . T . . . . . . . . E . . . . M O .
N . U . . . . . . . . . R . . . O C N I
. . A N O P E R O N . . . . . . P . . C N I
. . N . . . . . . . . . . . . E . . I .
. S O M A T I C . . . . . . . . R . .
```

8: ORIGINS OF LIFE AND THEORIES OF EVOLUTION: ANSWER KEY

BASIC TERMS CROSSWORD PUZZLE

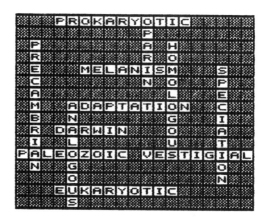

ADVANCED TERMS CROSSWORD PUZZLE

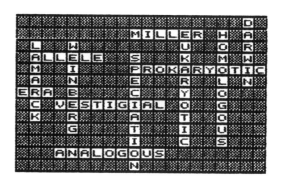

VOCABULARY WORDSEARCH

```
. . . . . . A . . . . . . . N I W R A D .
. . . . . L A N . . . . . . . . . . . . .
. . . . . L D O . . . . . . . . . . . . .
. . . . H E A I A N A L O G O U S . . . .
. . . . O L P T U . . . . . . . . . . . .
. . G . . M E T U . N H S . . . . . . . .
. P E . . O . A L . O A P . . . . . . .
. R O . . L . T O . I R E . P . . . L .
. E L . . O . I V . T D C . A . . . A .
. C O . . G . O E . A Y I . L . . . I .
C A G . . O . N . . R W A . E . . . G .
I M B . N U S . . . G E T . O . . . T .
O B C I R . . . . I I I . Z . . . S .
Z R T I A P . . . M N O . I . . . E V .
O A M P O . . . . . B . N . C . . V R E
N E O K . . . . . . E . . I . . . . .
C . . K C R A M A L G . . . . A . . .
. . . . . . . . . . . . . . . . . . . .
. . . . . . . . . . . . . . . . . . . .
```

9: PROTISTS: ANSWER KEY

BASIC TERMS CROSSWORD PUZZLE

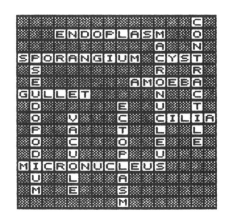

ADVANCED TERMS CROSSWORD PUZZLE

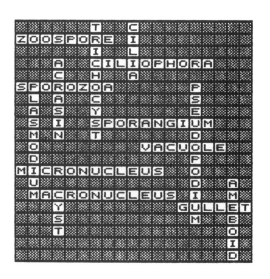

VOCABULARY WORDSEARCH

```
A . . C . . . A O Z O R O P S .
. M M Y X O M Y C O P H Y T A M .
. . O . . N . . . . . V . A . E
. . . E . P T . T . . A . C . L
. M . . B . S . R . S . . C . R . C
. S . . . A E P . A . Y . U . O . I
. A . . . . U A L . C . C O . U . L
. L . . E . D N . A . T . L . U . L
. P . T . O I . . S . I E . C . P
. O . . A B P D . . M . L . L . P
. T . L I O O P . . . O . E E .
. C . L N D C R . . . . D . U .
. E . E A I R O . . . . . I S . A
. . . G R U A T . . . . . U . I
. . . A Y M S I . . . . . . M . L
. . . . L . . S . . . . . . . I
. . . . F . . . T . E T A I L I C
. . . . . . . A . . . . . . . . .
. . . . . . . . M U I C E M A R A P .
```

10: AUTOTROPHIC PROTISTS: ANSWER KEY

BASIC TERMS CROSSWORD PUZZLE

ADVANCED TERMS CROSSWORD PUZZLE

VOCABULARY WORDSEARCH

```
. . . . . . A R Y G O R I P S . . . .
. . . . . . . . A L G I N A T E .
. . . . . . A T Y H P O E A H P . . .
. . . . . . . N O T K N A L P . .
. . . . . . A T Y H P O R O L H C
. . . . . M U I D I R E H T N A . .
. N . M . . E T Y H P O R O P S . .
. O . U . . M O T A I D . . . . .
. T B I O L U M I N E S C E N C E .
. K . N . . . . . . . . . . . . .
. N O . . . . E T Y H P O T E M A G
. A . G . . . A L G A E . . . . .
. L P . E . . . . . . . . . . . .
. P O . H . . . . . . . . . . . .
. O T . C . S U O T N E M A L I F . .
. T Y . R . . R A L U L L E C I N U .
. Y H . A . . . . . . . . . . . .
. H P . . . . . E T Y H P I P E . . . .
. P
```

11: FUNGI: ANSWER KEY

BASIC TERMS CROSSWORD PUZZLE

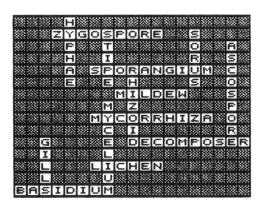

```
        H
    ZYGOSPORE       S
    P       T       O       A
    H       I       R       S
    A   SPORANGIUM          C
    E       E   H       S   O
            MILDEW          S
            M   Z           P
        MYCORRHIZA          O
        C   I               R
    G       E   DECOMPOSER
    I       L
    L   LICHEN
    L       U
BASIDIUM
```

ADVANCED TERMS CROSSWORD PUZZLE

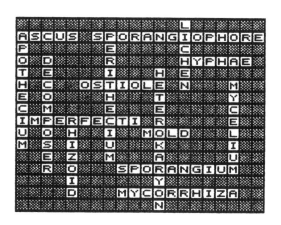

```
                        L
ASCUS   SPORANGIOPHORE
P           E       C
O   D       R       HYPHAE
T   E       I   H   E
H   C   OSTIOLE N       M
E   O       H   T       Y
C   M       E   E       C
IMPERFECTI  R       E
U   O   H   I   MOLD     L
M   S   I   U   K       I
    E   Z   M   A       U
    R   O   SPORANGIUM
        I       Y
        D   MYCORRHIZA
                N
```

VOCABULARY WORDSEARCH

```
. . M . . . . . . . . . . . . . . .
. . U . . . . . M U I L E C Y M H . .
. . S . . . . . . . . . . . E . . . .
. . H . E R O P S O G Y Z T . . . . .
. . R . E . . . . . M . E . . . . . .
. . O . P . . . O . R . . . . M . . .
. . O . I C S A L . O . E H . L U . .
. R M . T . . D . K . . T Y . . I . .
. E . . S . . A . . . I P . I D . . .
. S . . . . R S . . . S H . G I . . .
N O . . . . Y R C . . A A . . S . . .
E P . . O . H O A . R E . . A . . . .
H M . . N . I M F I . A . . M B . . .
C O . . T . Z Y U . G P . I . . . . .
I C . . S . O C N . N L . . . . . . .
L E . A . . I O G . D A . . . . . . .
. D E . . . D T I . E . . R . . . . .
. Y . . . . . A . W . . . O . . . . .
. . . . . . O S T I O L E . . P . . .
. . . M Y C O R R H I Z A . . . . S .
```

12: SEED PLANTS: ANSWER KEY

BASIC TERMS CROSSWORD PUZZLE

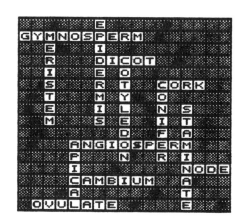

```
                E
GYMNOSPERM
        E       I
        R   DICOT
        I       E   O
        S   R   T   CORK
        T   M   Y   O
        E   I   L   N   S
        M   S   E   I   T
            D   F   A
        ANGIOSPERM
        P   N   R   I
        I           NODE
        CAMBIUM     A
        R           T
    OVULATE         E
```

ADVANCED TERMS CROSSWORD PUZZLE

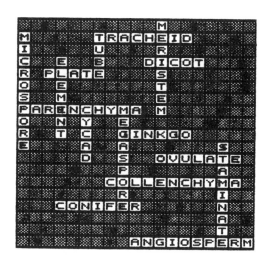

```
                M
M       TRACHEID
I       U       R
C   E   B   DICOT
R   PLATE       S
O   E   E       T
S   M           E
PARENCHYMA  M
O   N   Y   E
R   T   C   GINKGO
E       A   A       S
        D   S   OVULATE
            P       A
        COLLENCHYMA
        R           I
    CONIFER         N
                    A
                    T
            ANGIOSPERM
```

VOCABULARY WORDSEARCH

```
. A M Y H C N E R A P . . . . . . . .
. . . . O . M E T S I R E M G . . . .
. . . D D A C Y C . . . . . Y . E . .
. N . E . . . N E L L O P . M . M . .
. O . . D I E H C A R T . . M N . B .
. D . . . . . . . . . . . . I O . R .
D E L . . . . . . . . . . . C R . Y .
I C Y . S T A M I N A T E . R S . O .
C O T . . . . . . . . . . . O P . . .
O T O . A N G I O S P E R M . E R . .
T . C . . . . . . . . . . . S R . . .
. . . . . . . . . . . . . . P M . . .
. . . . . . . . . . . . . . O . . . .
. . . . . . S I M R E D I P E . R . .
. . . . . M O N O C O T . . . . E . .
. . . . . R E F I N O C . . . . . . .
. . . . . . . O G K N I G . . . . . .
. . . . . . . . . . . . . . . . . . .
. . . . . . . . . . . . . . . . . . .
. . . . . . C A M B I U M . . . . . .
```

13: ROOTS: ANSWER KEY

BASIC TERMS CROSSWORD PUZZLE

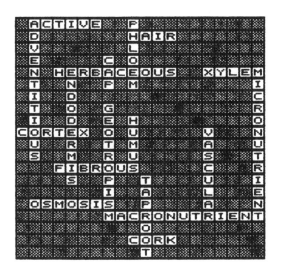

ADVANCED TERMS CROSSWORD PUZZLE

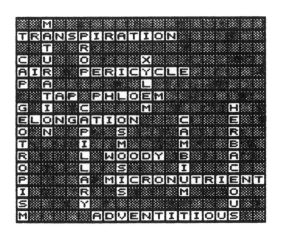

VOCABULARY WORDSEARCH

```
.  . C A P I L L A R Y . . . . . . . . . . .
.  . . . . S . . . C . T N A M R O D . . . .
E  . C . . . U . . O . . . . . . A . . . . .
L  . A . . . R . . R . . N . U . . O . . . .
O  . M . . R B . M E . . O X . . . O S . . .
N  . B . . S I . I X . . I . . . . M O . . .
G  . I P G U F . C . . . T . . . . . O S . .
A  . U E I O E . R . . . A . . . . . S I . .
T  . M R B E . . O . . . R . . . . . I S . .
I  . . I B C . . N . H . P . . . . L . S . .
O  . . C E A . . U . U . S . . . . A I . . .
N  . . Y R B . . T . M . . . . . . I R E . .
.  . . C E R . . R . U . . . . . . R E A . .
.  . M L L H . . I . S . . . . . . E A . . .
.  . E . I . . . N . M E O L H P . . . . . .
.  . L . N . . . T . . . . . . . . . . . . .
.  . Y . . . . T R O P I S M . . . . . . . .
.  . X . . . . . . . . . . . . . . . . . . .
```

14: STEMS: ANSWER KEY

BASIC TERMS CROSSWORD PUZZLE

ADVANCED TERMS CROSSWORD PUZZLE

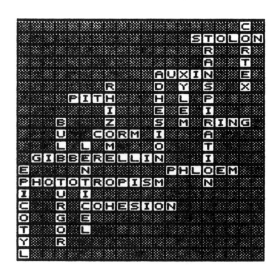

VOCABULARY WORDSEARCH

```
.  . . . . S I M R E D I P E . . . . P . . .
.  . . . . . R . . . . . . . . . . H . . . .
.  . . . . . O . . . . . . . . O . . . . N .
.  N . R . . G . . . . . . . T . . . . . O .
T  . U . O . . . . . . . T . . O . . . . I .
L  . . I . . S . . . O . . . . . . . C . T .
Y  . . . . E . . H . P . . . . . A . . . A .
T  . . . . . H I . . . . . . . M . . . . R .
O  . . . M . A . . . B . . . . . . . . . S .
C  A R . . C . . . . . . I . . . . . . . N .
I  U . H . . O . . . U . . . P . . . . . A .
P  E . . I . . R . . M . . . P H L . . . R .
E  . N . . . Z . T M . . . E . . O . . . T .
.  . . . . . O . . E . L . . X . E . . . . .
.  . . . . . . M . L . . . . M . . . . . . .
.  . . . . . . . E . Y . . . . . . . . . . .
.  . . . . . . . . X . . . . . . . . . . . .
```

193

15: LEAVES: ANSWER KEY

16: FLOWERS: ANSWER KEY

BASIC TERMS CROSSWORD PUZZLE

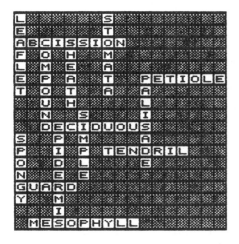

BASIC TERMS CROSSWORD PUZZLE

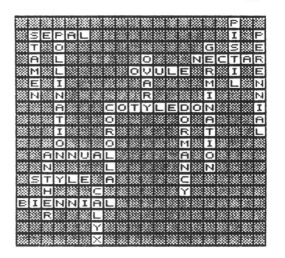

ADVANCED TERMS CROSSWORD PUZZLE

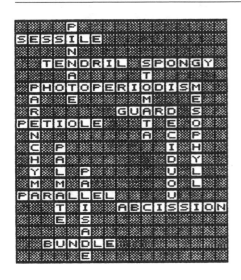

ADVANCED TERMS CROSSWORD PUZZLE

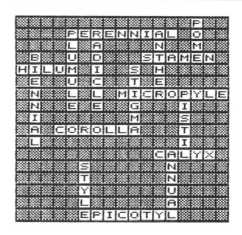

VOCABULARY WORDSEARCH

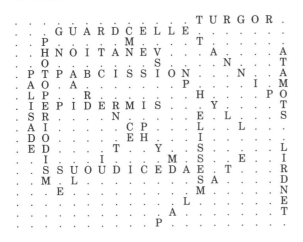

VOCABULARY WORDSEARCH

194

17: SPONGES AND COELENTERATES: ANSWER KEY

BASIC TERMS CROSSWORD PUZZLE

Answers visible in grid: BUDDING, AMEBOCYTE, ASYMMETRICAL, COLLAR, MOTILE, ECTODERM, SESSILE, VENTRAL

ADVANCED TERMS CROSSWORD PUZZLE

Answers visible in grid: SPONGIN, BLASTULA, SPICULE, METAZOAN, GASTRULA, POSTERIOR, DORSAL

VOCABULARY WORDSEARCH

```
. A . . . . . . . . . . . . . . . . . . . . . .
. N . F L A G E L L U M E L U M M E G .
. T E . . . . P O R I F E R A . . D .
E R . A M E B O C Y T E . . . . R . . D .
L I E R O P O T S A L B . . . . O . R .
I O R . S . C . . . . . . N . . I . R S . A L
S R . S . O . . . . . A . O . . R . A .
S . G P . L . . . . T . Y . O . E . L .
E . A O . L . . . . S . Z . T . T .
S . S N . A . . . . Y . A . E . S . P .
. . T G . R . . . . C . T . E . O .
. . R I G C . . . S . O . E . P .
. . U N N E . . . P . T E M .
. . L I L . . . . I . A . L .
. . A . D L . . . C . M . . I . . .
. . . . D . . . . U . E . N . . T .
. . . . U . . . . L . N . . . . O .
. . . . B . . . . E . . . . . . . M
```

18: ROTIFERS, FLATWORMS, ROUNDWORMS, AND SEGMENTED WORMS: ANSWER KEY

BASIC TERMS CROSSWORD PUZZLE

Answers visible in grid: GANGLION, FLAME, ESOPHAGUS, ANNELIDA, CYST, CLOSED

ADVANCED TERMS CROSSWORD PUZZLE

Answers visible in grid: VESICLE, CLITELLUM, CUTICLE, PORE, TURBELLARIA, GANGLION

VOCABULARY WORDSEARCH

```
. . . A . . M U L L E T I L C . . . .
. . . . N . . . . . . E . . . . .
. . . A . N . . . . E . . . . . .
C . . I . . E . . . L C I . . . . P
U . . R . . . L . . I . T . . . . R
T . . A N . . . I . D . S E . S . O
I . . L O . . . D . A . E . . Y . G
C . . L I T . N . A . E . V L . E C L
L . . E L G P . . . P . V . A . K U O
E . . B R N E . . H . . R N . L F T
. . R U A W O . . . . N I . F . I
. A T G O R M . . . . M D . . . D
. Q U . . M . . . . E . I . . .
. A D O T A M E R T . S . U .
. T . . . N E M A T O D E M .
. I S C O L E X C O E L O M . . .
. C . . . . . . . . . . . . . . .
```

19: MOLLUSKS AND ECHINODERMS: ANSWER KEY

BASIC TERMS CROSSWORD PUZZLE

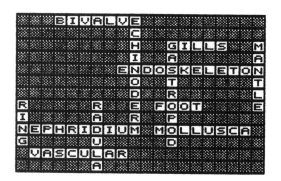

ADVANCED TERMS CROSSWORD PUZZLE

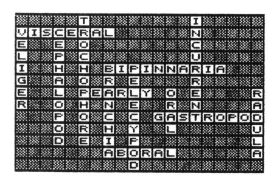

VOCABULARY WORDSEARCH

```
.  .  .  .  .  .  .  .  .  .  D  O  P  O  R  T  S  A  G  .  .
.  .  .  .  .  .  .  .  .  .  .  .  .  .  .  .  .  .  .  T  .
L  .  .  .  A  .  .  R  .  .  .  .  .  .  .  .  .  .  .  N  .
L  .  .  L  .  .  .  A  .  .  .  .  .  .  .  .  .  .  R  E  .
I  .  U  .  .  .  .  L  .  .  .  .  .  .  .  .  .  R  E  G  .
G  .  D  .  .  .  .  U  .  .  .  .  .  .  .  .  R  R  U  C  .
.  A  .  .  .  .  .  C  .  .  .  .  .  .  .  G  I  L  E  V  .
R  .  T  .  .  E  .  S  .  .  .  .  .  .  L  E  V  .  .  N  I
.  .  N  L  .  C  .  A  C  P  .  .  .  .  L  E  V  .  .  C  N
.  .  E  A  .  H  .  V  T  E  E  .  .  .  .  .  .  .  V  I  .
.  .  R  R  .  I  .  R  U  B  .  P  L  .  .  .  .  .  .  .  .
.  .  R  O  .  N  .  E  B  .  H  E  .  .  .  .  .  .  .  .  .
.  .  U  B  .  O  .  T  E  .  .  A  C  .  .  .  .  .  .  .  .
.  .  C  A  .  D  .  A  F  .  .  .  L  Y  .  .  .  .  .  .  .
.  .  X  .  L  E  .  W  O  .  .  .  .  O  P  .  .  .  .  .  .
.  .  E  .  L  R  .  O  .  .  .  .  .  .  P  O  .  .  .  .  .
.  .  .  A  .  M  .  .  T  .  .  .  .  .  .  O  D  .  .  .  .
.  .  R  .  .  .  .  .  T  E  V  L  A  V  I  B  .  .  D  .  .
.  O  M  O  L  L  U  S  C  A  .  .  .  .  .  .  .  .  .  .  .
```

20: ARTHROPODS: ANSWER KEY

BASIC TERMS CROSSWORD PUZZLE

ADVANCED TERMS CROSSWORD PUZZLE

VOCABULARY WORDSEARCH

```
.  .  .  .  .  .  .  N  O  I  T  A  T  N  E  M  G  E  S
.  .  .  .  .  .  .  .  .  .  .  N  I  T  I  H  C  .  .
.  E  L  B  I  D  N  A  M  .  .  .  .  .  .  .  .  .  .
.  .  .  .  .  .  .  X  .  .  .  .  .  .  .  .  .  .  .
.  .  .  T  .  .  A  A  R  A  C  H  N  I  D  .  .  .  .
.  N  .  R  .  C  .  R  .  .  D  .  .  S  .  .  .  .  .
.  O  N  A  .  N  A  .  O  .  O  .  .  T  .  .  .  .  .
.  T  E  C  .  O  R  .  H  .  P  .  .  A  .  .  .  .  .
.  E  M  H  .  S  A  .  T  .  O  R  .  T  .  A  .  .  .
.  L  O  E  .  S  L  P  .  O  .  U  .  O  .  N  .  .  .
.  E  D  A  .  E  A  .  L  .  R  .  .  C  .  T  .  .  .
.  K  .  B  .  T  C  .  A  .  .  .  .  Y  .  E  .  .  .
.  S  O  .  .  .  E  .  H  .  .  .  .  S  .  N  .  .  .
.  .  O  .  .  .  .  .  P  .  .  .  .  T  .  N  .  .  .
.  .  X  .  .  .  .  .  E  C  .  I  S  O  P  O  D  A  .
.  .  E  .  .  .  .  .  C  .  .  .  .  .  .  .  .  .  .
.  M  O  L  T  I  N  G  .  .  .  .  .  .  .  .  .  .  .
.  .  N  A  U  P  L  I  U  S  .  .  .  .  .  .  .  .  .
```

21: VERTEBRATES: ANSWER KEY

BASIC TERMS CROSSWORD PUZZLE

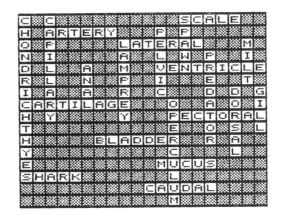

```
      INSTINCT
            O
VERTEBRATE              E
      U     O           N
IMMUNE I    C           D
   I        H           O
   C   ENDOSKELETON     C
DORSAL    A   R         C
      T   T   D         R
    REFLEX          GILL
                       N
      REPRODUCTIVE
```

ADVANCED TERMS CROSSWORD PUZZLE

```
C PECTORAL
H E
O L  N                    C
N V  O    CONDITIONED
D I  T         S          P
UROCHORDATA  AGNATHA
 I   C    E          A   E
 C   H    I    DORSAL    N
 H   O    C          O   D
 T   R    H          C   O
 H   D    T          O   S
 Y       AMPHIOXUS   R   K
 E       Y           R   E
 S   LEARNED         D   L
         S       INNATE
                     T   T
                     A   O
                         N
```

VOCABULARY WORDSEARCH

```
  . S . L . A .
U . E . A
 . R . Y . R . ENDOSKELETON .
 . O . H . O .           R .
    . C . T . TI         R E . D
    . . H . HNC . E      S . R R
    . O . O S C . E      P . O A
       . STR . I . P     O . H L
       . I T . D . R     N . C U
       . N . . E . A . D S . O C
    . C . . . I . T . N . E . T O
  . T A . . . C . A . O .    T I D
  . T A .      H . H         O N
  . A G N A T H A . T . E . C . . E P
  . D R          . H . T     Y . P P
  . O . . YRATNEMUGETNIA     A . A
  . H C . .                S . . N
  . C . . .                   . I
```

22: FISHES: ANSWER KEY

BASIC TERMS CROSSWORD PUZZLE

```
C  C              SCALE
HARTERY      F  F
O  F     LATERAL        M
N  I     A  L W F       I
O  L  A  M VENTRICLE
R  L  R  F I  E  T
I  A  A  R C  D  O  G
CARTILAGE     O  A  O  I
H  Y     Y  PECTORAL
T        E     O  S  L
H    BLADDER   R  A
Y        C     L
E        MUCUS
SHARK    L
      CAUDAL
         M
```

ADVANCED TERMS CROSSWORD PUZZLE

```
 F C  I
 E R  N        E
OLFACTORY      C
 V N  E        T
 I I  G  CYCLOSTOMATA
 C A  U        T
   L  M        H
   O  ENDOTHERMIC
 SPAWN         R         V
 E   T CHROMATOPHORE
 R        I           I
 C   OSTEICHTHYES     N
 U   F
MILT ATRIUM
 U   I
 M   C
```

VOCABULARY WORDSEARCH

```
      . . C . NWAPS . . . .
E . . CY . . . . . . . L
RO . PECTORALS . Y . . LA
OH . L . . . . CY . . AT
PO . MO . . O . AR . . RE
OT . LLS . . S . LO . C . RA       C
TA . UO . . T . ET . HO . AL       I
AM . CMG . I . A . L . LI          MRE
MO . RAT . ACL . F . DR . NE       HTO
RH . DETA . LHA . L . RIC . E      TOT
HC . OPA . ITM . O . HC            OTE
C . S . LRYR . TH . P . . CH       TLADUAC
 . A . I . AEE . . T . H           . . CE
 . L . M . CSY . . H . Y
 . ENDOTHERMICYE
 .              S GILL . .
 . ANAL . . .
```

197

23: AMPHIBIANS: ANSWER KEY

BASIC TERMS CROSSWORD PUZZLE

```
              C
    A  S      ALIMENTARY
TYMPANIC   O
    P  L      AMPLEXUS
    H  A   N  C
    I  M   METAMORPHOSIS
    B  A   W        O
    I  NICTITATING  N
    A  O             E
    N  EUSTACHIAN    Y
    R
```

ADVANCED TERMS CROSSWORD PUZZLE

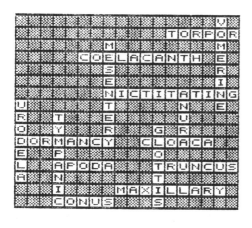

```
                          V
                   TORPOR
         M              E
      COELACANTH       R
         S             I
      NICTITATING
U      T        N   E
R   T  E        U
O   Y  R        G   R
DORMANCY    CLOACA
E   F        Q
L  APODA    TRUNCUS
A   N        T
    I   MAXILLARY
    CONUS       S
```

VOCABULARY WORDSEARCH

```
   . A N U R A
. T Y M P A N I C
      R
      O       A D O P A   . M   Y   G
 . P Y R A L L I X A M . E   C N   N
  R   Y  A         T     A M R   I T
 O. Y   L       A       M  R O   T A
T   R   E       M        O N  C O  T I
    E   D     O   R   A   L  D   . T C
    T   O   R   P   I   O       . C  I
    N   R   U H   B   A         . N N
    E   O   I   C             . N
    S   S   H   A
  M  I   P
    S   M
        A
```

24: REPTILES: ANSWER KEY

BASIC TERMS CROSSWORD PUZZLE

```
ALLIGATOR           C
      M             R
      N  T    M     O
TORTOISE      O     C
      U  O  REPTILIA O
VIPER T  R        T  D
  G   T  E     I  G  I
  U   L  P  INTERNAL
  A   E  I     G  C  E
  N      N        K
CARAPACE     DINOSAUR
```

ADVANCED TERMS CROSSWORD PUZZLE

```
O
VIVIPAROUS      N
O     A     Q  HEMOTOXIN
V     R     U  U
I     I   PLASTRON
V  E  E C  M    O
I  C  T H  A  T   A
F  D  A O  T  O   L
A  Y  L R  A  X   L
R  S  I    C  I   A
OVIPAROUS  AMNION
U  S    N     I    T
S            M     O
             A  SKINK
             N      S
```

VOCABULARY WORDSEARCH

```
        E C A P A R A C           S
                                  I
  S                               S Y D
  U  O V O V I V I P A R O U S . D C
  O        S Q U A M A T A    .   C E
  R . N E U R O T O X I N . P
  A                        A R
  P    H E M O T O X I N . I   S
  I V            M        E   U
  I V            O        T   O
  V             L        A   R
                T        L   A
                I            P
                N            I
                G            V
                             O
```

BASIC TERMS CROSSWORD PUZZLE

```
  Q
  U         I
M I G R A T I O N
  L     A     C L O A C A
  L Y O L K   U           C
        O   B           A
A L B U M E N     F O L L I C L E
        T     E       R   C
      G I Z Z A R D   O   U
        Q     T   P   M
      A   N   H
      V       E
      E N D O T H E R M Y
      S
```

ADVANCED TERMS CROSSWORD PUZZLE

```
              A   B
    T   P     C   L A
    H A R C H A E O P T E R Y X
V A N E   E     N   R   B
    C   C       T   I   U
    O   D O W N   O   C   L
    D   C         U   I   E
    R A C H I S   R   A
    N       N     A   L
    T       L
```

VOCABULARY WORDSEARCH

```
. . . . E . . . . . . . . . . . . . . . .
. . . . N . . C . . . . . . . L . . . . .
. . . . . D . O . . . . . . . A I . . . .
. . X . . . O N D O W N . . . I C . . . .
P . Y . . . T . O H R A C H I S I R . . .
. R . R . . U . . E . . R . . . T L . . .
. E . R . . R . . . R . . M . . T A . . .
. . C M . T . P . . . . M . . . L A . . .
V . A . O . L C . O . T H E C O D O N T .
. A . N . L C . I . E . . . . . A . . . .
. . N . E I . A . A . . . H . . . . . . .
. . T I . N G . . C . L . . . . . . . . .
. . N G O . R . . . . . C . . . . K . . .
. . . O . . . . . . . . R . . L . . . . .
. . . P . . . . . . . . . . A O . . . . .
. . . . . . . . . . . . . . . Y . . . . .
```

BASIC TERMS CROSSWORD PUZZLE

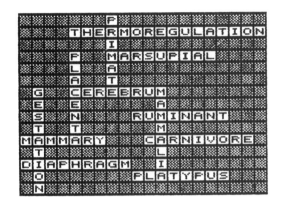

```
              P
      T H E R M O R E G U L A T I O N
              I
      P   M A R S U P I A L
      L   A
      A   T
  G   C E R E B R U M
  E   E         A
  S   N     R U M I N A N T
  T   T     M
M A M M A R Y   C A R N I V O R E
  T           L
D I A P H R A G M   I
  O           F L A T Y P U S
  N
```

ADVANCED TERMS CROSSWORD PUZZLE

```
P       M A R S U P I A L   R
R       R             O
I N S E C T I V O R A   D
M       E             E
A       O   C A R N I V O R A
T       S   E       T
E       C   T
        E R U M I N A N T
        N O   C
        T O   E
P L A C E N T A   A
  T   N
  A   T
  I
      L A G O M O R P H A
```

VOCABULARY WORDSEARCH

```
. . O . . . . . . . . . . . . . . . .
. . M . . L P . . . . . . . . . . . .
. . N . A R O V I N R A C . . . . I S
. . I . I I M . . . . . . . N . S U .
. . V . P U M . . N . . . S . E . P Y
. . O . U A T . O . . P C . A . . T A
. . R . S R E I . L T . A E . . . A L
. . E . R A . T A . A I . I D . . L P
. . . M A . T . A . H . C V . T I . .
. . . T . P . E O . N C . . . U N . .
. . S . R . N R . E S . . . . . N A .
. . E . O . T A . D O . . . . G . N I
. G . M . A . . O B . . . . . U . U M
. . O . . . . R O . . . . . L A T U R
. G . . . . . R . . . . . . . A T E .
. A . . . . P . . . . . . . . T E C .
L . . . . . . . . . . . . . . . . . .
. . . . . . . . . . A E C A T E C . .
```

BASIC TERMS CROSSWORD PUZZLE

```
H        H
U     B  O
M   C I  M
ANTHROPOLOGY
N   I  E        N
O   M  O  G     E
I   F  MAGNON   A
D   A  L  O     N
N      O        D
Z      HALFLIFE R
SAPIENS    L    T
    E      L    H
            PRIMATE
                L
```

ADVANCED TERMS CROSSWORD PUZZLE

```
                  H
                  U   M
ANTHROPOMORPHISM  A
U                 A   G
SOCIOBIOLOGY      N   N
T             H   O   O
R           SAPIENS
A   BIPEDAL       L   D
L   O             F
O   R   T         L
P   A   A         I
I   D   S         F
T   I   I         E
H   O   I
ERECTUS     M
C   A   M
U   R
S   B
    O
NEANDERTHAL
```

VOCABULARY WORDSEARCH

```
                              S  P
R                          O  R
A       L     C     S    C  I        L
D     A   H   U  O  T  A              A H
I       D   I  B  T  C        E        H T
O       E   I  M  E        R           T R
C   M   P O   P       R              E E
A   U   L I         A        N         D N
B   I   O   B           Z             A E
O   S  G             E             N
N   A  HOMOSAPIENS  E           A E
    T                              N
    O
    P
         ANTHROPOLOGY
DIONAMUH
```

BASIC TERMS CROSSWORD PUZZLE

```
APPENDICULAR
T  X          X      YELLOW
R  T     CARDIAC        R
O  RED   A       A      I
P  N     R   CLAVICLE   G
H  S     T   R          I
Y  O     I   A   LIGAMENT
   R     L   N         N
         S   I         S
      S  A   G E
      M    E FLEXOR
      O         T
      OSSIFICATION
      T  T       O
      H  R       N
         I
PERICARDIAL
      T
      E
      D
```

ADVANCED TERMS CROSSWORD PUZZLE

```
        P
        E
THORACIC
   I    M
   OSTEOCYTE
   S    O
   T  MYOFIBRIL
   E    I      U  A
   U    L      M  D
   M    A      B  D
        M      A  U
        U  E CERVICAL
        ACTIN     T
        T  T      O
        O     SUTURE
SACRUM
```

VOCABULARY WORDSEARCH

```
              P
              E
              L   M                D
           L  V   Y  S             I
           A  I   S  O   F       A
   L       I  X   I  F   E     P
L  I    C  A   B  R   N    H   R
I  G    A  I   R  U   A      R  O   A
G  A    I  V   I  M  G       S  N   D
A  O  A B   R  L  M          N  E   D
M  E  D  O   E          C    E  T   U
E  N  O           REDMARROW  X  C   C
N  T  M                      E  T   T
T  I                         PERIOSTEUM O
   N  A                            R
   A  L
   L
```

© 1991 by Center for Applied Research in Education

29: HUMAN BODY PLAN—DIGESTIVE SYSTEM: ANSWER KEY

BASIC TERMS CROSSWORD PUZZLE

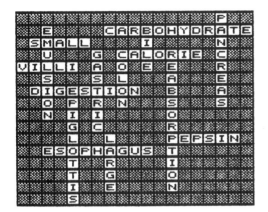

ADVANCED TERMS CROSSWORD PUZZLE

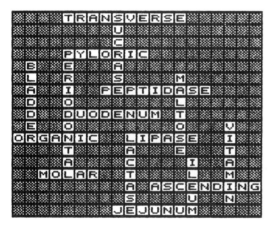

VOCABULARY WORDSEARCH

```
 .  .  .  .  .  .  .  .  .  .  .  .  .  .  .  .  .  .  .
 .  .  .  .  .  G  N  I  D  N  E  C  S  E  D  .  .  .  .
 .  .  .  .  .  .  L  I  .  .  .  .  .  .  .  .  .  .  .
 .  .  .  .  .  .  I  P  .  .  .  .  .  .  .  E  .  P  .
 .  .  .  .  .  .  P  A  .  .  .  .  S  .  Y  .  .  .
 .  E  S  A  T  C  A  L  .  .  R  .  L  .  .  .  .  .
 .  V  I  T  A  M  I  N  .  E  .  V  .  R  .  O  .  .  .
 .  .  .  .  .  S  C  I  .  .  .  .  .  .  .  .  .
 .  .  .  .  N  A  C  .  .  .  .  .  .  .  .  .  .
 .  .  .  A  L  .  .  .  .  .  .  .  .  .  .  .  .
 P  A  N  C  R  E  A  S  R  O  .  .  .  .  I  .  .
 .  .  .  .  T  R  .  .  .  L  .  .  .  .  .  .
 .  .  .  I  .  .  .  E  .  .  .  .  .  .  .
 .  .  E  .  .  .  U  .  .  .  .  .  .  .
 .  .  .  .  .  M  .  .  .  .  .  .  .  .
 .  .  .  .  .  E  T  A  L  A  P  .  .  .  .
 .  .  E  T  A  R  D  Y  H  O  B  R  A  C  .
 .  .  C  I  R  T  S  A  G  .  .  .  .  .
```

30: HUMAN BODY PLAN—CIRCULATORY AND EXCRETORY SYSTEMS: ANSWER KEY

BASIC TERMS CROSSWORD PUZZLE

ADVANCED TERMS CROSSWORD PUZZLE

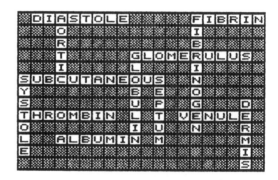

VOCABULARY WORDSEARCH

```
 .  .  .  .  .  .  .  .  .  .  A  O  R  T  I  C  .  .  .
 .  .  .  .  .  .  .  .  .  S  E  P  T  U  M  .  .
 .  .  .  .  .  .  N  E  P  H  R  O  N  .  .  .
 T  .  .  .  .  .  .  .  .  .  .  .  .  .  .  .
 .  H  .  M  U  I  D  R  A  C  I  R  E  P  .  .  .  .
 .  R  .  Y  .  .  .  .  .  .  .  .  .  .  .
 .  O  .  R  .  .  L  .  .  .  .  .  .  .  .  .
 .  .  M  .  E  .  A  .  .  .  .  .  .  .  .
 .  .  B  .  T  N  .  C  O  R  O  N  A  R  Y  .  N  .
 .  .  .  I  E  R  .  .  .  .  .  .  .  .  .  N  .
 .  .  .  R  N  .  A  .  .  .  .  .  .  .  I  .
 N  I  E  V  .  .  .  .  .  .  .  .  .  .  R  .
 .  .  .  .  .  .  .  .  .  .  .  .  .  .  B  .
 .  .  .  .  .  P  U  L  M  O  N  A  R  Y  .  F  .
 .  .  .  .  .  .  .  .  .  .  .  .  .  .  .
 .  .  E  T  A  N  I  T  U  L  G  G  A  .  .
```

BASIC TERMS CROSSWORD PUZZLE

ADVANCED TERMS CROSSWORD PUZZLE

VOCABULARY WORDSEARCH

BASIC TERMS CROSSWORD PUZZLE

ADVANCED TERMS CROSSWORD PUZZLE

VOCABULARY WORDSEARCH

© 1991 by Center for Applied Research in Education

33: HUMAN BODY PLAN—ENDOCRINE SYSTEM: ANSWER KEY

BASIC TERMS CROSSWORD PUZZLE

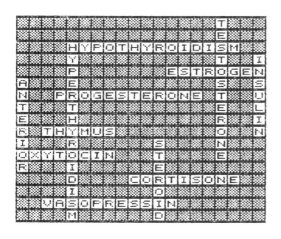

```
                    H     ADRENAL
                    O     O
                    M   CORTISONE
        ESTROGEN    O     E
                    N     N
        PANCREAS  DIABETES
        I         T     L   N
        T         A     I   D
        U      INSULIN   O
        I      G   I     E   C
        T      O   S     B   R
        A      N         I
        R      A       GLAND
THYROID                  E
```

ADVANCED TERMS CROSSWORD PUZZLE

```
                          T
                          E
        HYPOTHYROIDISM
        Y             T   I
        P         ESTROGEN
A       E                 S   S
N       PROGESTERONE  T   U
T       T         R   E   L
R  THYMUS         S   R   I
I       R         S       O   N
OXYTOCIN          T       E
R       I         E
        D       CORTISONE
        I         D
        VASOPRESSIN
        M         D
```

VOCABULARY WORDSEARCH

```
.  .  .  .  .  .  .  .  .  .  .  .  .  .  .  .  .  P
.  .  .  .  .  .  .  .  .  E  N  I  R  C  O  X  E  .  R  .
P  .  .  .  .  .  .  .  .  .  .  .  .  .  .  .  .  O  H
I  S  E  T  E  B  A  I  D  .  .  .  .  .  .  .  G  H  .  .
T  .  .  E  N  D  O  C  R  I  N  E  .  .  E  E  Y  .  .
U  .  V  .  .  A  .  .  .  .  .  N  S  P  .  .  .
I  .  .  A  .  .  .  .  .  .  O  T  O  .  .  .
T  .  .  .  S  .  .  .  .  D  S  E  T  .  .  .
A  .  .  .  .  O  .  .  .  I  R  H  .  .  .
R  .  .  .  .  .  P  .  .  T  O  Y  .  .  .  E
Y  .  N  .  .  P  .  T  O  Y  .  .  .  N
.  .  I  L  .  .  .  R  R  N  R  E  .  .  I  L
.  .  L  .  C  .  .  O  E  O  E  S  .  .  A  N
.  .  U  .  .  D  .  .  S  I  S  T  .  .  L  A  E
.  .  S  .  I  .  .  .  .  I  .  .  .  N  R  D
.  .  N  .  S  .  .  .  .  .  N  .  .  E  R  D  A
.  I  M  .  .  .  O  X  Y  T  O  C  I  N  R  D  A
                                             A
```

34: HUMAN BODY PLAN—REPRODUCTIVE SYSTEM: ANSWER KEY

BASIC TERMS CROSSWORD PUZZLE

```
            E     P     M
P         P I     E     E
L         I S     I     N
A         O T     M     S
C         D   OVULATION   T
E     N   Y   G   A   R   E
N     M   E   N   UTERUS
FALLOPIAN     T   A     T
E     U   S       A   T   I
T     T           T   I   C
U     EMBRYO      I   O   L
S     U           N       E
AMNION            N
```

ADVANCED TERMS CROSSWORD PUZZLE

```
      PROSTATE
R     E               O   F
O     M       ZYGOTE  E   E
G     I               F   T
E     N         UTERUS
S     I         R   R   S
T     F   T     E   E
E     E   E         T
OVARY R   S         H   S
O     O   T     R
N     UMBILICAL
E     S   C
            PLACENTA
            E
```

VOCABULARY WORDSEARCH

```
.  .  .  .  .  .  .  .  P  .  .  .  .  .  .  .
.  .  .  .  .  .  L  .  .  .  N  .  .  .  .  .
.  .  .  .  .  A  .  .  .  .  O  .  .  .  .  .
.  .  .  C  A  M  .  P  .  .  I  .  .  .  .  .
.  .  E  .  N  N  .  R  .  .  T  .  .  .  .  .
.  .  N  .  E  I  .  O  .  .  A  .  .  .  .  .
A  .  .  .  G  O  .  G  .  .  N  .  .  .  .  .
.  .  .  .  O  N  .  E  .  .  A  L  .  .  .  .
.  .  .  .  R  .  .  S  .  .  P  .  .  .  .  .
E  .  .  .  T  .  E  .  M  .  R  .  M  .  U  .
.  T  .  .  S  .  E  .  R  .  E  I  .  T  .  .
.  .  O  .  .  .  .  .  O  .  P  .  E  .  .  .
.  .  .  G  .  .  .  .  N  .  S  .  R  S  U  T  E  F  .
.  .  .  .  Y  .  .  .  .  .  .  U  .  .  .  .
.  .  .  .  .  Z  .  .  .  .  .  S  .  .  .  .
.  .  .  L  U  T  E  I  N  I  Z  I  N  G  .  .  .
.  .  O  V  U  L  A  T  I  O  N  .  .  .  .  .
.  .  F  E  R  T  I  L  I  Z  A  T  I  O  N  .  .  .  .
```

35: TOBACCO, ALCOHOL, AND OTHER DRUGS: ANSWER KEY

BASIC TERMS CROSSWORD PUZZLE 1

```
HALLUCINOGEN
           M  R
           UPPER
ADDICTION  H  F
M          Y  L
P          S  E
H    COCAINE  X
E       I  M
T       R  A
BARBITURATE
M       H
I NARCOTIC
N       S
E       I
  DEPRESSANT
```

BASIC TERMS CROSSWORD PUZZLE 2

```
         STIMULANT
         R
CODEINE VALIUM
      A  N
     DRUG Q
O    C  U   M
PSYCHOACTIVE O D
I    T  L   R G
U  NICOTINE P W
M    C  Z   H N
        SEDATIVE
        R   N R
            E
```

VOCABULARY WORDSEARCH

36: INTRODUCTION TO ECOLOGY: ANSWER KEY

BASIC TERMS CROSSWORD PUZZLE

```
L        M
INTERTIDAL
M        R
I  ABIOTIC
T  I     N
I BIOSPHERE
N I  M
G O  ECOLOGY
  T  H
  I  A
  CONIFEROUS
     N
```

ADVANCED TERMS CROSSWORD PUZZLE

```
            BIOSPHERE
                    C
       XEROPHYTE    O
N          E        L
INTERTIDAL     B    O
T          A  UPWELLING
R  F    L  G   N    Y
O  L  NITRIFICATION
G  A  T  C     H
E  N  T     MARINE
N  K  Q        C
  TERRESTRIAL
  Q  A
  N CLIMAX
```

VOCABULARY WORDSEARCH